P9-EDU-857

AUGUSTANA UNIVERSITY COLLEGE
LIBRARY

From
Sea Charts
to
Satellite Images

From
Sea Charts
to
Satellite Images

Interpreting North American
History through Maps

EDITED BY
DAVID BUISSERET

THE UNIVERSITY OF CHICAGO PRESS
CHICAGO AND LONDON

The following plates are from the Newberry Library:
1.1, 1.3, 1.4, 1.6, 1.8, 1.9, 1.11, 2.4, 2.5, 2.5a, 2.7,
2.8, 2.9, 3.1, 3.3, 3.6, 3.7, 5.3, 5.4, 5.5, 5.6, 5.7a,
5.8, 5.8a, 5.9, 5.9a, 5.10, 5.10a, 6.5, 6.6, 6.7, 6.8,
7.1, 7.2, 7.4, 7.5, 7.6, 7.7, 8.6, 8.9, 8.10, 8.11,
9.1, 9.2, 9.3, 9.4, 9.5, 9.8, 9.9, 9.10, 10.3, 10.4,
11.1, 11.2, 11.3, 11.6, 12.3a, color plates 1, 3, 4,
and 5.

The University of Chicago Press,
Chicago 60637
The University of Chicago Press, Ltd.,
London
© 1990 by The University of Chicago
All rights reserved. Published 1990
Printed in the United States of America

99 98 97 96 95 94 93 92 91 90   5 4 3 2 1

Library of Congress Cataloging in Publication Data

From sea charts to satellite images : interpreting North American
 history through maps / edited by David Buisseret.
   p. cm.
   Includes bibliographical references.
   ISBN 0-226-07991-0 (alk. paper); ISBN 0-226-07992-9 (pbk.; alk. paper)
   1. United States—History—Maps. 2. North America—History—Maps.
 3. Cartography—History. I. Buisseret, David.
 E179.F84 1990   68968
 973′.22′3—dc20   89-20521
                       CIP

♾ The paper used in this publication
meets the minimum requirements of the
American National Standard for
Information Sciences—Permanence of
Paper for Printed Library Materials,
ANSI Z39.48-1984.

Contents

AUGUSTANA UNIVERSITY COLLEGE
LIBRARY

List of Plates

Chapter 6
Bird's-Eye Views of Towns and Cities

Chapter 7
City Maps and Plans

Chapter 8
North American County Maps and Atlases

Preface

All historical events, and indeed all historical developments, have taken place in both time and space. If we want to understand these events and developments, it is surely as important to locate them in space as to situate them in time. For spatial location, the map is the essential instrument, but it seems that maps are often neglected by historians, particularly in the United States. Conscious of this apparent weakness, the National Endowment for the Humanities in 1980 and again in 1982 funded Summer Institutes in Cartography at the Newberry Library, pilot projects aimed at working out how the thirty faculty participants could best use maps in their teaching and research. Some attention was given to the drawing of new maps to illustrate social phenomena, but the institutes mostly concentrated on developing ways of using "old maps" in humanistic studies. The sixty participants in the two institutes seemed to find them useful, and indeed developed many new map-based courses upon their return to their institutions.

These proceedings have encouraged us to try to spread some of these ideas more widely. In 1984 the National Endowment for the Humanities agreed to fund the commissioning of eight experts to write chapters on map types with which they were familiar, and these pieces form the core of this book. The contributions were of course originally in somewhat disparate shape, but with the unrelenting and highly constructive aid of Anne Knowles they have been licked into what I hope is a format common enough for the reader to see what we are trying to do. Each chapter begins with a general survey of the nature and history of the type of map in question, goes on to give eight or ten commentated examples (usually with explanatory maps) of the way such maps can be used by historians, and concludes with a section on how to find this material.

We have tried to cover many periods and places, and have consequently taken much advice along the way. We owe particular thanks to Debra Mancoff, Amy Myers, and Angela Miller for reading the chapter on landscape views, to Priscilla Strain for looking at the chapter on aerial images, and to Rainer Vollmar and James Akerman for general criticisms. Of the contributors, Gerald Danzer and Robert Karrow read widely outside their own chapters. Marsha Sellmer, of the Map Library at the University of Illinois (Chicago), lent us much material, and Ken Cain, photographer at the Newberry Library, carried out many delicate copying jobs with great success. Finally, Tom Willcockson drew the maps and diagrams, but also entered fully into questions of substance in the text, making many useful suggestions.

I do not doubt that, even with all this help, I have missed some of the contributors' mistakes and even introduced some of my own into their work. But I hope that these errors will not be so conspicuous as to spoil the pleasure of readers, for this is a book not only to look things up in, but also, we hope, to savor and enjoy.

DAVID BUISSERET

Contributors

CHRISTOPHER BARUTH is map and imagery librarian at the American Geographical Society Library in Wisconsin-Milwaukee, and is responsible for looking after a large collection of aerial images.

DAVID BUISSERET is director of the Hermon Dunlap Smith Center for the History of Cartography at the Newberry Library, Chicago, and has long been interested in using maps as historical sources. See, for example, *Historic Jamaica from the Air,* which he wrote with Jack Tyndale-Briscoe (Barbados, 1969). He is also the author of *Historic Illinois from the Air,* forthcoming from the University of Chicago Press.

MICHAEL P. CONZEN, professor of geography at the University of Chicago, edited *Chicago Mapmakers* (Chicago, 1984), and is author in particular of "The County Landownership Map in America," *Imago Mundi* 36 (1984): 9–31. He is working on a book-length treatment of this subject.

GERALD DANZER is professor of history at the University of Illinois at Chicago. He has often used cartographic material in his publications, of which the most recent is *Public Places: Exploring Their History* (Nashville, 1987).

LOUIS DEVORSEY is professor of geography at the University of Georgia. He has been particularly active in using historical maps as evidence in determining both national and international boundaries, as in his most recent book, *The Georgia-South Carolina Boundary* (Athens, Ga., 1982).

RONALD E. GRIM is head of the reference and bibliography section of the Geography and Maps Division at the Library of Congress. He has written several studies using the maps of the General Land Office as historical evidence, and is the author of *Historical Geography in the United States: A Guide to Information Sources* (Detroit, 1982).

J. B. HARLEY was formerly professor of geography at the University of Exeter, in England, and now holds the same title at the University of Wisconsin-Milwaukee. He is co-editor of the six-volume *History of Cartography,* now in the course of publication, and has written extensively on the use of maps in English local history, in such works as *The Ordnance Survey and Land-Use Mapping* (Norwich, 1979).

ROBERT KARROW is curator of maps at the Newberry Library. He edited the *Checklist of Printed Maps of the Middle West to 1900* (14 vols., Boston, 1981–83), and has extensive experience in helping readers find maps to throw light on historical problems.

DAVID B. QUINN was formerly professor of modern history at the University of Liverpool, and is the author of a great many books and articles concerning English exploration in the New World. His most recent book on this theme is *Set Fair for Roanoke* (Chapel Hill, N.C., 1985).

THOMAS SCHLERETH is professor of American studies at the University of Notre Dame. He is the author of *Artifacts and the American Past* (Nashville, 1980), and has frequently used maps of twentieth-century and earlier dates in elucidating problems of North American history.

From
Sea Charts
to
Satellite Images

Text and Contexts in the Interpretation of Early Maps

J. B. Harley

*Old maps are slippery witnesses. But where would
historians be without them?*

J. H. Parry, 1976

*A*mong the many classes of documents regularly used by historians, maps are well known but less well understood. We could compile an anthology of statements that categorize maps not only as "slippery" (the adjective used by the distinguished historian J. H. Parry), but also as "dangerous" or "unreliable." Historians have tended to relegate maps—along with paintings, photographs, and other nonverbal sources—to a lower division of evidence than the written word (Rotberg and Rabb 1988). Much historical research and writing is undertaken without systematic recourse to contemporary maps. Moreover, even where maps are admitted as documents, they are regarded as useful principally for a narrow range of selected historical questions. It is widely acknowledged, for instance, that maps are valuable for such topics in United States history as its discovery, exploration, territorial expansion, and town planning. Less frequently are they considered as offering crucial insights into processes of social history. When a historian reaches for a map, it is usually to answer a fairly narrow question about location or topography and less often to illuminate

cultural history or the social values of a particular period or place. Why should maps have suffered such neglect?

Part of the answer, as already noted, lies in the attitudes of historians. Writing about the history of maps *per se* has been at best a marginal interest for mainstream historians: when, we may ask, did an article about cartography last appear in *The American Historical Review?* Yet part of the problem also lies with those who call themselves historians of cartography. In describing the bibliographical and technical complexity of maps, they have failed to communicate an understanding of their social nature. In the light of these tendencies, the answer to the question "What is a map?" is a vital preliminary to the fruitful interrogation of maps as historical documents.

MIRROR OR TEXT?

The usual perception of the nature of maps is that they are a mirror, a graphic representation, of some aspect of the real world. The definitions set out in various dictionaries and glossaries of cartography (Wallis and Robinson 1987) confirm this view. Within the constraints of survey techniques, the skill of the cartographer, and the code of conventional signs, the role of a map is to present a factual statement about geographical reality. Although cartographers write about the art as well as the science of mapmaking, science has overshadowed the competition between the two approaches. The corollary is that when historians assess maps, their interpretive strategies are molded by this idea of what maps are claimed to be. In our own Western culture, at least since the Enlightenment, cartography has been defined as a factual science. The premise is that a map should offer a transparent window on the world. A good map is an accurate map. Where a map fails to deal with reality adequately on a factual scale, it gets a black mark. Maps are ranked according to their correspondence with topographical truth. Inaccuracy, we are told, is a cartographic crime.

This value judgment is often translated into the way we read old maps. It promotes a mode of interpretation that emphasizes the factual or literal statements maps make about an empirical reality. Whether depicting the Caribbean landfall of a sixteenth-century navigator or the relict features of some ghost town from a nineteenth-century mining boom, the map is judged in terms of the positioning of its coordinates, the shape of its outlines, or the reliability of features measured in the landscape. It is used purely and simply as a quarry of facts in the reconstruction of the past. I am not suggesting that we downgrade this historical application of old maps. As an index to the location of things, processes, and events in the past maps are a unique form of documentation. Locating human actions in space remains the greatest intellectual achievement of the map as a form of knowledge.

There is, however, an alternative answer to the question "What is a map?" For historians an equally appropriate definition of a map is "a social construction of the world expressed through the medium of cartography." Far from holding up a simple mirror of nature that is true or false, maps redescribe the world—like any other document—in terms of relations of power and of cultural practices, preferences, and priorities. What we read on a map is as much related to an invisible social world and to ideology as it is to phenomena seen and measured in the landscape. Maps always show more than an unmediated sum of a set of techniques. The apparent duplicity of maps—their "slipperiness"—is not some idiosyncratic deviation from an illusory perfect map. Rather it lies at the heart of cartographic representation. Herein lies a historical opportunity. The fascination of maps as humanly created documents is found not merely in the extent to which they are objective or accurate. It also lies in their inherent ambivalence and in our ability to tease out new meanings, hidden agendas, and contrasting world views from between the lines on the image.

In introducing ways of interpreting the maps of America described in this volume, I propose a different interpretative metaphor. They will be discussed as text rather than as a mirror of nature. Maps are text in the same senses that other nonverbal sign systems—paintings, prints, theater, films, television, music—are texts. Maps also share many common concerns with the study of the book, exhibiting a textual function in the world and being "subject to bibliographical control, interpretation, and historical analysis" (McKenzie 1986). Maps are a graphic language to be decoded. They are a construction of reality, images laden with intentions and consequences that can be studied in the societies of their time. Like books, they are also the products of both individual minds and the wider cultural values in particular societies.

Signs, Symbols, and Rhetoric

Like all other texts, maps use signs to represent the world. When these become fixed in a map genre, we define them as conventional signs. Maps do not possess a grammar in the mode of written language, but they are nonetheless deliberately designed texts, created by the application of principles and techniques, and developed as formal systems of communication by mapmakers. In modern cartography strenuous efforts have been made to standardize these rules of map composition. Textbooks and models (Robinson et al. 1984) tell us how the world should "best" be graphically represented in terms of lines, colors, symbols, and typography. For some of the older maps that are described below there were also rule books for their construction and design, and vocabularies of different signs. Such works can act as a grammar or dictionary in learning to read or translate the map text.

The symbolic dimension of maps also links them to other texts. Modern cartographers usually regard their maps as factual statements written in the language of mathematics, but they are always metaphors or symbols of the world. A mode of interpreting such symbolic layers of meaning by employing iconographical principles will be discussed below (pp. 11–13).

Maps are also inherently rhetorical images. It is commonplace to say that cartography is an art of persuasion. What goes against modern wisdom is to suggest that *all* maps are rhetorical. Today's mapmakers distinguish maps that are impartial or objective from other maps used for propaganda or advertising that become "rhetorical" in a pejorative sense. Cartographers also concede that they employ rhetorical devices in the form of embellishment or ornament, but they maintain that beneath this cosmetic skin is always the bedrock of truthful science. What I am suggesting is that rhetoric permeates all layers of the map. As images of the world, maps are never neutral or value free or ever completely scientific. Each map argues its own particular case. The thematic maps discussed in chapter 9, for example, are especially rhetorical. They are part of a persuasive dis-

course, and they intend to convince. Theirs is not an innocent reality dictated by the intrinsic truth of the data; they are engaging in the ancient art of rhetoric. Most maps speak to targeted audiences, and most employ invocations of authority, especially those produced by government, and they appeal to readerships in different ways. The study of the history of cartographic representation, when employed as an aid to the interpretation of maps as historical documents, is also a history of the use of the different rhetorical codes employed by mapmakers (Wood and Fels 1986).

THE CARTOGRAPHER'S CONTEXT

The basic rule of historical method is that documents can only be interpreted in their context. The rule applies equally to maps, which must be returned to the past and situated squarely in their proper period and place. The readers of this book may be disappointed to learn how little contextualization of maps there is in the literature of the history of cartography. Connoisseurs' books on maps, for example, are oblivious of the social reality beyond the decorative price tag. Technical specialists in the history of maps, those trained as cartographers, seldom step beyond the workshop door and into the outside world. Context is simplistically portrayed as "general historical background." What is lacking is a grasp of context as a complex set of interactive forces—a dialogue with the text—in which context is central to the interpretative strategy. We tend to regard context as "out there" and the maps we are studying as "inside." Only when we knock down this barrier—this false dichotomy between an externalist and an internalist approach to historical interpretation—can map and context be studied in an undivided terrain. To achieve this it is necessary to distinguish between three aspects of context (La Capra 1983) that intersect the reading of maps as texts. The three aspects of context in my argument will be (1) the context of the cartographer, (2) the contexts of other maps, and (3) the context of society.

The context of the cartographer is best represented in the literature of early map interpretation. It is almost sixty years since the historian J. A. Williamson wrote, "It is impossible to be dogmatic about the evidence of maps unless we know more than we commonly do about the intention and circumstances of those who drew them" (Williamson 1929). This simple dictum—enshrining the why, who, and how approach to maps—is a good starting point. Yet the relationship between the maker and map is far from straightforward. It is neither a simple question of establishing authorship—as with books and documents—nor of determining the intention of the mapmaker.

With respect to authorship, if we exclude manuscript maps that are unambiguously identified and have a known provenance, the historian is frequently confronted with disentangling multiple authorship. Most maps are the product of a division of labor. As we enter the long transition from the manuscript age to the age of printing, the cartographic division of labor is accentuated, the author becomes a shadowy figure, and the translation from mapped reality to map is more complex. The question arises, "To what extent was a particular map the work of a surveyor, an editor, a draftsman, or an engraver?" Who has determined its form and content? As we concern ourselves with different craftsmen, Williamson's question about circumstances becomes more difficult to resolve. The relationship between the facts of the mapmakers' lives and what appears in the map is correspondingly fragmented. Within the frame of one map there may be several texts—"an intertextuality"—that has to be uncovered in the interpretative process.

More than many other texts, maps are thus mediated by a series of technical activities, each performed by a different "author." R. A. Skelton once wrote: "As bibliography to literary criticism, or as diplomatic to the interpretation of medieval documents, so is the technical analysis of early maps to the studies they serve" (Skelton 1965). It is this requirement—reconstructing the technical contexts of mapmaking—that places a heavy demand on the ancillary skills of the historian. The student of early maps may have to become an expert on the histories of different types of maps (Wallis and Robinson 1987), be well versed in navigation and surveying techniques (Singer et al. 1954–78), be familiar with the processes by which maps were compiled, drafted, engraved, printed, or colored, and know something about the practices of the book and map trades. Every map is the product of several processes involving different individuals, techniques, and tools (Woodward 1974). To understand them, we need to deploy specialist knowledge from subjects as diverse as bibliography and paleography, the history of geometry and magnetic declinations, the development of artistic conventions, emblems and heraldry, and the physical properties of paper and watermarks. The pertinent literature is likewise scattered in a large number of disciplines and modern languages (Skelton 1972; Harley 1987), straddling the history of science and the history of technology as well as the humanities and social sciences. But how the author or authors of a map made it in a technical sense is always a first step in interpretation.

Establishing the map maker's intention is similarly less straightforward than might appear at first sight. Every map codifies more than one perspective on the world. As an expression of intention, function remains a key to reading historical maps, but such purposes were often loosely defined, or the map was directed at more than one kind of user. While we may accept, for example, that fire-insurance maps (Chapter 9) have a single use, many of the other groups of maps described in this volume were designed for a variety of purposes. Such multiple aims complicate the assessment of maps as historical documents. Topographical maps (Chapters 3 and 10), or city maps and plans (Chapter 7), were made to fulfill several needs at once. They were designed as administrative or jurisdictional records, for defense, for economic development, or perhaps as general works of topographical reference. The simple link between function and content breaks down. It is inadequate, for example, to define a topographical survey as merely producing a "map showing detailed features of the landscape." Topographical map series were often of military origin, and they emphasized features of strategic significance. In the United States, even after the Geological Survey assumed control of the national topographical survey in 1879, maps were still expected to serve logistical military purposes as well as geological and other civilian functions. Even today we can detect traces of the military mind in the woodland density categories of USGS maps that are still classified in relation to the ease with which infantry can move through the countryside (Thompson 1981). In many nineteenth-century topographical maps (Chapter 10), with military needs in mind, relief was similarly emphasized at the expense of cultural detail.

Intention thus cannot be fully reconstructed through the actions of individual mapmakers. A simple intention may still be found in individual manuscript maps, but there are also broader aspects of human agency that impinge on interpretation. Cartographic intention was seldom merely a question of an individual's training, skill, available instruments, or of the time and money needed to complete the job properly. Cartographers were rarely independent decision makers or free of financial, military, or political constraints. Above the workshop there is always the patron, and consequently the map is imbued with social as well as technical dimensions. We might do well to adapt to cartography the words of Michael Baxandall on fifteenth-century Italian painting. Such art was always

the deposit of a social relationship. One one side there was a painter who made the picture, or at least supervised its making. On the other side

there was somebody else who asked him to make it, provided funds for him to make it and, after he had made it, reckoned on using it in some way or other. Both parties worked within institutions and conventions—commercial, religious, perceptual, in the widest sense social—that were different from ours and influenced the forms of what they together made. (Baxandall 1972).

In much of history, the cartographer was a puppet dressed in a technical language, but the strings were pulled by others.

The role of patronage varies considerably in the maps of America. With earlier manuscript maps, such as those of the age of European exploration (Chapter 2), patrons were powerful individuals—kings or queens, princes or popes. By the nineteenth century, however, American map makers were increasingly dragooned by larger institutions such as the General Land Office (Chapter 4) and the United States Geological Survey (Chapter 10). Personal map-making skills were subordinated not only to sets of standard instructions designed to make whole classes of maps uniform (Chapter 4), but also to state and federal politics. With political influence in mind, we should be chary of interpreting the official topographic surveys of the United States as "standard" historical documents. It has been said that

the geodetic and topographic surveys conducted by the federal government throughout the nineteenth century evolved as byproducts of ad hoc Congressional legislation and the personal intervention of civil servants, and not as a result of a national policy for mapmaking (Edney 1986).

Both the geographical order in which surveys were conducted and the content of the maps were influenced by the need to map first areas with valuable mineral deposits. Policy concerns as much as the skills of individual mapmakers gave rise to the diverse images of the American landscape preserved in the national series of topographical maps.

In qualifying the limits of the individual cartographer's influence, I am not denying that "mapmakers are human" (Wright 1942). Unusual personal skill as well as idiosyncrasy still flourishes in the interstices of institutional practice. In the maps of the township and range system, for example, "possibilities of error, omission, personal bias and even misrepresentations abounded" (Chapter 4). Even in today's machine generated maps and aerial images (Chapter 12), historians should remain alert to deviant ways in which individual technicians may have inscribed their routine tasks. This may be more difficult to detect behind the assertive rhetoric of computer technology, but again the standard historical record does not exist.

Similar observations may be made about commercial mapping. This forms an important part of the cartographic historical record in the United States (Ristow 1985), but it also shows conflicts of interest. The market place usually constrains the free play of cartographic standards. One text we always read in these maps is a financial balance sheet. "Where the detective hunts for fingerprints," it has been remarked, "we must look for profit if we are to understand the basic mechanism of early map publishing. . . . No salesman ever tells the whole truth and it would be an unwary historian who took land sale maps for a true cartographic record" (Campbell 1989). Moreover, as the size of map businesses increases and print runs grow longer, cartography acquires a corporate image. The patron is now a larger public or perhaps a special interest group, such as the consumers of highway maps, who look over the cartographer's shoulder to influence what is being mapped.

THE CONTEXT OF OTHER MAPS

A major interpretative question to be asked of any map concerns its relationship to other maps. The inquiry has to be focused in different ways. For example, we could ask: (1) What is the relationship of the content of a single map (or some feature within it) to other contemporary maps of the same area? (2) What is the relationship of such a map to maps by the same cartographer or map-producing agency? (3) What is its relationship to other maps in the same cartographic genre (of one bird's-eye view, for instance, to other North American bird's-eye views)? (4) Or what is the relationship of a map to the wider cartographic output of an age? The questions vary but their importance is universal. No map is hermetically closed upon itself nor can it answer all the questions it raises. Sooner or later early map interpretation becomes an exercise in comparative cartography (Harley 1968). The cartographic characteristics of the larger family may enable anonymous maps to be identified, unusual signs or conventions interpreted, or inferences made about the parameters of accuracy. Our confidence in a map document may be increased (or diminished) when it exhibits the proven characteristics of a larger group.

In this part of contextual study a corpus of related maps is built around the single map. Just as in the analysis of literary texts the unity or identity of a corpus of texts has to be constructed (La Capra 1983), so too in early map interpretation we can follow definite procedures. These can be applied to a group of maps of the same period but, equally, the depiction of an area or feature can be traced on a series of maps through time. Three approaches will be noted below and they may be used either separately or in combination in evaluating a single map within the larger group.

The comparative study of linear topographical features on maps (such as coastlines, river networks, or a system of trails and highways) is a well-tried technique. Outlines are reduced to common scale and are then compared visually. Examples appear in the classic nineteenth-century studies of early maps (Nordenskiöld 1889) and the method can also be adapted to the digital analysis of linear features by computer (Morrison 1975). A recent application of the older method is to the Spanish and French mapping of the Gulf of Mexico in the sixteenth and seventeenth centuries (Buisseret 1987). After "photocopying, assembling, and examining a great many maps" (p. 4) it was possible, on the basis of salient features in coastal outlines, to identify five main phases of mapmaking. Through the use of this comparative classification, individual maps were then assigned to stages of development and their origin, sources, and topographical reliability were assessed from the characteristics of the larger group.

But if every map has a genetic fingerprint that the method helps to identify, caution must also be exercised. The study of outlines may fall short of providing conclusive evidence of provenance. There are many pitfalls. R. A. Skelton has written that "visual impressions suggesting affinity or development of the outline in two maps may be misleading if we do not take into account the licence in drawing or interpretation that the cartographer might allow" (Skelton 1965). Or again, there may be technical variations influencing the shape of map outlines or their graticules of latitude and longitude. Maps are easily corrupted in the process of copying, or they may derive from surveying or navigation techniques that have been obscured in the process of compilation. Before the nineteenth century, maps were frequently aligned toward magnetic rather than true north. Magnetic declination varied locally and changed through time so that map makers were unable, in the absence of systematic observations, to correct for this factor. It remains a critical source of error in the comparison of outlines (Lanman 1987; Skelton 1965).

A second aspect of the comparative analysis of early maps involves the study of place-names or toponymy. Like outlines, place-names offer a way of constructing genealogies and source profiles for previously scattered maps. Indeed, the two methods are often used in conjunction, as in the classic studies of the early cartography of the Atlantic coast of Canada (Ganong 1964). Yet the cross-tabulation of the names on a series of maps as a means of classification or of establishing the interrelationships of the group must also be approached with caution (Richardson 1984a and b). In initial periods of exploration, Europeans of different nationality would have heard names from the mouths of native American speakers of a variety of languages, and they would have attempted to record them in accordance with their own sound system, in far from standardized spellings. Even where European names were applied to North American geography, there was ample scope for corruption in the processes of translating and editing them: the names attest to carelessness, misreading, or misunderstanding by successive generations of cartographers who had no first hand knowledge of the places or languages involved. Of names on the maps of the sixteenth-century Dieppe school of cartographers (Wallis 1981), for example, it is said that "no two Dieppe cartographers coincide completely in the number of names they record, while spelling varies widely and even the positioning of names is not always consistent" (Richardson 1984a). Not surprisingly, place-names have sometimes been used uncritically for purposes of comparing maps (Skelton 1965). The sound practice is to confine the analysis to only those names unambiguously common to a number of maps.

The third method for comparative cartography—carto-bibliography—has the largest literature. Not only have the definition and finer points of the method been extensively discussed (Karrow 1976; Verner 1965, 1974), but its practice is fully represented in a series of fundamental works on early American cartography (Winsor 1884–89; Wheat 1957–63; Cumming 1958). The aim of carto-bibliography is to bring together a series of maps printed from the same printing surface. It applies equally to the woodcut, copperplate, lithographic, or other map-printing processes (Woodward, 1975). By this method a sequence of geographical and other changes in related maps can be reconstructed. This in turn allows the publication history of the maps of an area to be pieced together. It also allows the single map to be dated and slotted into this sequence, and the extent of geographical revision between states or editions of maps to be detected. Maps are often representations of time as much as space. As Skelton puts it, we discover how "matter from various horizons of time or intellectual development" is incorporated into their images. And we learn that "the search for the ultimate source may lead us back through many stages of revision or adaptation, derivation or transcription, compilation" (Skelton 1965, 28–29). Carto-bibliography is thus a basic tool of the map historian. Either as a technique or as a means of measuring the channels and rate of diffusion of geographical knowledge (thereby linking maps to the context of society), its insights are indispensible.

THE CONTEXT OF SOCIETY

The third context of cartography is that of society. If the map maker is the individual agent, then society is the broader structure. Interpretation—reading the cartographic text—involves a dialogue between these two contexts. The framework of definite historical circumstances and conditions produces a map that is inescapably a social and cultural document. Every map is linked to the social order of a particular period and place. Every map is cultural because it manifests intellectual processes defined as artistic or scientific as they work to produce a distinctive type of knowledge. There is no neat causal arrow that flows from society into the map, but rather causal arrows that flow in both directions. Maps are not outside society: they are part of it as constitutive elements within the wider world. It is the web of interrelationships, stretching both inside and beyond the map document, that the historian attempts to read. In exploring this reflexivity, two strategies might be used to survey the context of society in the maps of America.

The Rules of Cartography

The first strategy is to attempt to identify "the rules of the social order" within the map (Foucault 1972). Every map manifests two sets of rules. First, there are the cartographers' rules, and we have seen how these operate in the technical practices of map making. The second set can be traced from society into the map, where they influence the categories of knowledge. The map becomes a "signifying system" through which "a social order is communicated, reproduced, experienced, and explored" (Williams 1982). Maps do not simply reproduce a topographical reality; they also interpret it.

The rules of the social order are sometimes visible, even self-evident, within a group of maps. Alternatively, they are sometimes hidden within the mode of representation. Among the category of "visible society," we may place the North American bird's-eye views of towns and cities (Chapter 6), the city maps and plans (Chapter 7), and the county maps and atlases. They are all cultural texts taking possession of the land (Clarke, 1988). All proclaim a social gospel and serve to reinforce it. Bird's-eye views of towns, for instance, "sing the national anthem of peace and prosperity, of movement and openness of calm and order, and of destinies to be fulfilled" (Chapter 6). The map wears its heart on its sleeve and it comes alive in a context of frontier ethics and patriotism as topography is decoded from the emphatically rhetorical style of the image.

Where the social rules of cartography are concealed from view, a hidden agenda has to be teased out from between the lines on the map. Such a map is duplicitous, and a different strategy is called for. Instead of picking up social messages that the map emphasizes, we must search for what it de-emphasizes; not so much what the map shows, as what it omits. Interpretation becomes a search for silences (Harley 1988), or it may be helpful to "deconstruct" the map (Harley 1989) to reveal how the social order creates tensions within its content. Among the maps that could be so elucidated are some of the eighteenth-century large-scale maps (Chapter 3), the topographic surveys of the United States (Chapter 10), and the aerial images (Chapter 12). Here technology has suppressed social relations. Because they appear to be accurate or objective, such maps are often viewed as nonproblematic documents. A satellite image or a topographical map made by "scientific" methods, so it is believed, has a moral and ethical neutrality. It is a factual and straightforward document. So long as we recognize *technical* limitation, the pathway of interpretation is secure.

Such assumptions are false. Representation is never neutral, and science is still a humanly constructed reality. The large-scale maps of eastern North America in the mid–eighteenth century (Chapter 3) illustrate this contention. At first sight they meet the goals of Enlightenment cartography. They are built on geodetic measurements; they begin to show "cartographic mastery" over the landscapes of eastern North America; and they suppress some of the more overtly fanciful, mythic, and pictorial elements of earlier maps. Take a closer look, however, and they also signal the territorial imperatives of an aggressive English overseas expansion (Harley forthcoming). Colonialism is first signposted in the map margins. Titles make increasing reference to empire and to the possession and bounding of territory; dedications define the social rank of colonial governors; and cartouches, with a parade of national flags, coats-of-arms, or crowns set above subservient Indians, define the power relations in colonial life (Clarke 1988). But the contours of colonial society can also be read between the lines of the maps. Cartography has become preeminently a record of colonial self-interest. It is an unconscious portrait of how successfully a

European colonial society had reproduced itself in the New World, and the maps grant reassurance to settlers by reproducing the symbolic authority and place-names of the Old World. Moreover, as the frontier moved west, the traces of an Indian past were dropped from the image. Many eighteenth-century map makers preferred blank spaces to a relict Indian geography (Harley forthcoming). I do not suggest that the omissions—the "rules of absence"—were deliberately enforced in the manner of a technical specification. But even where they were taken for granted, or only subconsciously implemented, to grasp them helps us to interrogate early maps.

The Meaning of Maps

Another interpretative strategy applies the iconographical methods of art history to maps. Iconography is defined as "that branch of the history of art which concerns itself with the subject matter or meaning of works of art" (Panofsky 1955). The question "What did the map mean to the society that first made and used it?" is of crucial interpretive importance. Maps become a source to reveal the philosophical, political, or religious outlook of a period, or what is sometimes called the spirit of the age. An iconographical interpretation can be used to complement the rules-of-society approach. While the latter reveals the tendencies of knowledge in maps—its hierarchies, inclusions, and exclusions—the former examines how the social rules were translated into the cartographic idiom in terms of signs, styles, and the expressive vocabularies of cartography.

The essence of iconographical analysis is that it seeks to uncover different layers of meaning within the image. Panofsky suggested that in any painting we encounter (1) a primary or natural subject matter consisting of individual artistic motifs; (2) a secondary or conventional subject matter that is defined in terms of the identity of the whole painting as a represen-

TABLE 1. ICONOGRAPHICAL PARALLELS IN ART AND CARTOGRAPHY

ART (Panofsky's terms are used)	CARTOGRAPHY (suggested cartographic parallel)
1. Primary or natural subject matter: artistic motifs	Individual conventional signs
2. Secondary or conventional subject matter	Topographical identity in maps: the specific place
3. Intrinsic meaning or content	Symbolic meaning in maps: ideologies of space

tation of a specific allegory or event (he gave the example of a painting of the Last Supper); and (3) a symbolic layer of meaning that often has an ideological connotation. This does not offer a neat formula for early map interpretation, but it may be ventured that the levels of meaning in a map are similar to those in a painting (Harley 1983). These parallel levels in the two forms of representation are summarized in Table 1.

First, at the level 1, the individual signs, symbols, or decorative emblems on a map are made equivalent to the individual artistic motifs. While the full meaning of any single sign may become apparent only when viewed in the mosaic of other signs in the map as a whole, for some interpretative purposes it may be necessary to evaluate the content and meaning of individual signs (for example, as well as establishing its cultural meaning, we may need to know how far the sign for the depiction of a church or a house on an early map is reliable from the architectural point of view).

Second, the identity of the real place represented on a map is assumed to be the equivalent of Panofsky's level 2 or second stage in interpretation. Its apprehension involves a recognition that a particular map is that of a plantation in South Carolina, of Boston, or of California. It is at this level—that of the real place—that maps have been most used by his-torians. Moreover, it is for evaluating the real places in maps that most interpretive techniques, whether devoted to their planimetric accuracy or to their content, have been developed. There are numerous exemplars for this type of topographical scholarship (Blakemore and Harley 1980).

The third interpretive level in a map is the symbolic stratum. Until recent years, apart from the contributions of a handful of art historians (Schulz 1978), this hermeneutic dimension of early cartography was neglected. Only recently has interpretation moved to embrace a symbolic and ideological reading of early maps. Here we accept that maps act as a visual metaphor for values enshrined in the places they represent. The maps of America described in this volume are always laden with such cultural values and significance, plotting a social topology with its own culturally asserted domain. Maps always represent more than a physical image of place. A town plan or bird's-eye view (Chapters 6 and 7) is a legible emblem or icon of community. It inscribes values on civic space, emphasizing the sites of religious belief, ceremony, pageant, ritual, and authority. Or in the nineteenth-century county and historical atlases (Chapter 8), there is more on the maps than an inert record of a vanished topography. What we read is a metaphorical discourse, as thick as any written text, about immigrant rural pride, about Utopias glimpsed, about order and prosperity in the landscape. Such maps praise possession of the land, enshrine property demarcations, and memorialize farm buildings and the names of property holders. Through both word and image they appealed to the industry and patriotism of the new Americans. And the longer we look the more symbolic cartography becomes. Thus a Rand McNally highway map (Chapter 11) speaks to the American love affair with the automobile, and even the seemingly earthy maps of the United States Geological Survey (Chapter 10) are a symbolic assertion of the changing perceptions and priorities of society rather than just maps of objects in the landscape. In such ways, "maps speak, albeit softly, of subtle value judgments" (Stilgoe 1983). To read the map properly the historian must always excavate beneath the terrain of its surface geography.

CONCLUSION

By accepting maps as fundamental documents for the study of American past, we begin to appreciate how frequently maps intersect major historical processes. From territorial treaties to town planning, and from railroads to the rectangular grid, they underlie the making of modern America (Stilgoe 1982). But if this is an immense practical contribution, neither should we ignore the historical influence of real maps upon the more elusive cognitive maps held by generations of Americans since the sixteenth century. In addition to regarding the map as a topographical source, we are becoming aware of a cartographic power that is embedded in its discourse (Boelhower 1988; Harley 1989). The power of the map, an act of control over the image of the world, is like the power of print in general (McLuhan 1962; Eisenstein 1979). Since the age of Columbus, maps have helped to create some of the most pervasive stereotypes of our world.

How the historian uses a map also depends on the context of the individual scholar. Insights are determined not only by the intrinsic qualities of a particular map but also by the historical investigation in hand, by its objectives, by its research methods, and by all the other evidence that can be brought to bear on its problems. Just as there are innumerable maps of America for the historian to consult, so there is an equally unlimited list of research topics for which maps may be appropriate. It has not been my intention to play down the technical aspects of early map interpretation, but in view of the fact that these already have an extensive literature, it seemed important to take this opportunity to sketch in a broader framework within which they can be deployed. The three contexts of cartography that have been outlined are never mutually exclusive but are subtly and often inextricably interwoven. Maps, once we learn how to read them, can become uniquely rewarding texts for the historian.

REFERENCES

Baxandall, Michael. 1972. *Painting and Experience in Fifteenth Century Italy: A Primer in the Social History of Pictorial Style.* Oxford: Clarendon Press.

Blakemore, M. J., and J. B. Harley. 1980. *Concepts in the History of Cartography: A Review and Perspective.* Monograph. *Cartographica* 17, no. 4.

Boelhower, William. 1988. "Inventing America: A Model of Cartographic Semiosis." *Word & Image* 4, no. 2 (April–June): 475–97.

Buisseret, David. 1987. "Spanish and French Mapping of the Gulf of Mexico in the Sixteenth and Seventeenth Centuries." In *The Mapping of the American Southwest,* ed. Denis Reinhartz and Charles C. Colley, 3–17. College Station, Texas: A & M University Press.

Campbell, Tony. 1989. "Knowledge and Market Mechanism as Impulses for Map Publishing." In *Abstracts, XIIIth International Conference on the History of Cartography,* 55–56. Amsterdam.

Clarke, G. N. G. 1988. "Taking Possession: The Cartouche as Cultural Text in Eighteenth-Century American Maps." *Word & Image* 4, no. 2 (April–June): 455–74.

Cumming, William P. 1958. *The Southeast in Early Maps*. Princeton: Princeton University Press; rev. ed. 1962.

Edney, Matthew H. 1986. "Politics, Science, and Government Mapping Policy in the United States, 1800–1925." *The American Cartographer* 13, no. 4: 295–306.

Eisenstein, E. L. 1979. *The Printing Press as an Agent of Change: Communications and Cultural Transformations in Early-Modern Europe*. Cambridge: Cambridge University Press.

Foucault, Michel. 1972. *The Archaeology of Knowledge and the Discourse on Language*. Translated from the French by A. M. Sheridan Smith. New York: Pantheon Books.

Ganong, W. F. 1964. *Crucial Maps in the Early Cartography and Place-Nomenclature of the Atlantic Coast of Canada*. With an Introduction, Commentary, and Map Notes by Theodore E. Layng. Toronto: University of Toronto Press.

Harley, J. B. 1968. "The Evaluation of Early Maps: Towards a Methodology." *Imago Mundi* 22: 62–74.

Harley, J. B. 1983. "Meaning and Ambiguity in Tudor Cartography." In *English Map-Making 1500–1650: Historical Essays,* ed. Sarah Tyacke, 22–45. London: British Museum Publications.

Harley, J. B. 1987. "The Map and the Development of the History of Cartography." In *The History of Cartography,* Vol. 1, *Cartography in Prehistoric, Ancient, and Medieval Europe and the Mediterranean,* ed. J. B. Harley and David Woodward, 1–42. Chicago: The University of Chicago Press.

Harley, J. B. 1988. "Silences and Secrecy: the Hidden Agenda of Cartography in Early Modern Europe." *Imago Mundi* 40:57–76.

Harley, J. B. 1989. "Deconstructing the Map." *Cartographica* 26, no. 2: 1–20.

Harley, J. B. Forthcoming. "Power and Legitimation in the English Geographical Atlases of the Eighteenth Century." In *Images of the World: The Atlas Through History,* ed. John A. Wolter. Washington, D.C.: Library of Congress.

Karrow, R. W. "Carto-bibliography." *AB Bookman's Yearbook,* Part 1, 43–52.

Lanman, Jonathan. 1987. *On the Origin of Portolan Charts*. Chicago: The Hermon Dunlap Smith Center, The Newberry Library.

La Capra, Dominick. 1983. *Rethinking Intellectual History: Texts, Contexts, Language*. Ithaca: Cornell University Press.

McKenzie, D. F. 1986. *Bibliography and the Sociology of Texts*. The Panizzi Lectures, 1985. London: The British Library.

McLuhan, Marshall. 1962. *The Gutenberg Galaxy: The Making of Typographic Man*. Toronto: University of Toronto Press.

Morrison, J. L. 1975. "Recommendations for the Classification of the Extent Maps of the Great Lakes." Unpublished report to the Hermon Dunlap Smith Center for the History of Cartography, Newberry Library, Chicago.

Muehrcke, Phillip C. 1986. *Map Use: Reading, Analysis, and Interpretation*. 2nd ed. Madison, Wisconsin: J P Publications.

Nordenskiöld, A. E. 1889. *Facsimile-Atlas to the Early History of Cartography with Reproductions of the Most Important Maps Printed in the XV and XVI Centuries*. Stockholm; New York: Dover Publications, 1973.

Panofsky, E. 1955. *Meaning in the Visual Arts*. New York.

Parry, J. H. 1976. "Old Maps are Slippery Witnesses." *Harvard Magazine*. Alumni ed. (April), 32–41.

Richardson, W. A. R. 1984a. "Jave-la-Grande: A Place Name Chart of Its East Coast." *The Great Circle* 6, no. 1: 1–23.

Richardson, W. A. R. 1984b. "Jave-la-Grande: A Case Study of Place-Name Corruption." *The Globe* 22: 9–32.

Ristow, Walter. 1985. *American Maps and Mapmakers: Commercial Cartography in the Nineteenth Century*. Detroit: Wayne State University Press.

Robinson, Arthur H., et al. 1984. *Elements of Cartography,* 5th ed. New York: John Wiley & Sons.

Rotberg, Robert I. and Theodore K. Rabb. 1988. *Art and History: Images and Their Meaning*. Cambridge: Cambridge University Press.

Schulz, J. 1978. "Jacopo de' Barbari's View of Venice: Map Making, City Views and Moralized Geography before the Year 1500." *Art Bulletin* 60, 425–74.

Singer, Charles, et al., eds. 1954–1978. *A History of Technology*. Vols. 1–4. Oxford: Clarendon Press.

Skelton, R. A. 1965. *Looking at an Early Map*. Lawrence, Kansas: University of Kansas Libraries.

Skelton, R. A. *Maps*. 1972. *A Historical Survey of Their Study and Collecting*. Chicago: University of Chicago Press.

Stilgoe, John R. 1982. *Common Landscape of America, 1580 to 1845*. New Haven and London: Yale University Press.

Stilgoe, John R. 1983. "Mapping Indiana: Nineteenth-Century School Book Views." In *Perceptions of the Landscape and Its Preservation*, by John R. Stilgoe, Roderick Nash, and Alfred Runte. Indiana Historical Society Lectures.

Thompson, Morris M. 1981. *Maps for America: Cartographic Products of the U.S. Geological Survey and Others*. 2nd ed. Reston, Va.: U.S. Department of the Interior.

Verner, C. 1965. "The Identification and Designation of Variants in the Study of Early Printed Maps." *Imago Mundi* 19: 100–105.

Verner, C. 1974. "Carto-bibliographical Description: The Analysis of Variants in Maps Printed from Copperplates," *The American Cartographer* 1, no. 1: 77–87.

Wallis, Helen, ed. 1981. *The Maps and Texts of the Boke of Idrography presented by Jean Rotz to Henry VIII*. Oxford: Roxburghe Club.

Wallis, Helen M., and Arthur H. Robinson. 1987. *Cartographical Innovations: An International Handbook of Mapping Terms to 1900*. Tring, Herts: Map Collector Publications.

Wheat, C. I. 1957–1963. *Mapping the Transmississippi West, 1540–1861*. 5 Vols. San Francisco: Institute of Historical Cartography.

Williams, Raymond. 1982. *The Sociology of Culture*. 1st American ed. New York: Schocken Books.

Williamson, J. A. 1929. *The Voyages of John and Sebastian Cabot*. London: The Argonaut Press.

Winsor, Justin. 1884–1889. *Narrative and Critical History of America*. 8 Vols. Boston: Houghton, Mifflin and Co.

Wood, Denis, and John Fels. 1986. "Designs on Signs: Myth and Meaning in Maps." *Cartographica* 23, no. 3: 54–103.

Woodward, David. 1974, "The Study of the History of Cartography: A Suggested Framework." *The American Cartographer* 1, no. 2: 101–115.

Woodward, David, ed. 1975. *Five Centuries of Map Printing*. Chicago: University of Chicago Press.

Wright, J. K. 1942. "Map Makers Are Human: Comments on the Subjective in Mapping." *Geographical Review* 32: 527–544.

CHAPTER ONE

The European Antecedents of New World Maps

David Buisseret

Almost all the people who came to North America used maps of one kind or another. Those who arrived by crossing the Bering Strait (the so-called *Indians*) developed ingenious methods for recording the surrounding topography, using such materials as lay to hand. In some lands such as Mexico, they established thriving and extensive cartography schools, whose products are abundant and whose history is slowly emerging. But because historians have barely begun to write about the cartography of North American Indians, we shall be concerned with only the maps that were drawn by Europeans, first as visitors and later as residents.

The sixteenth century, which is, roughly speaking, the first century of European exploration of North America, was a period of prodigious development in European cartography, and most of the new techniques and styles found a rapid application in the mapping of North America. Out of the European Middle Ages came one of the most distinctive of cartographic forms, the portolan chart. This type of map apparently emerged fully formed about 1200. For a long time portolan charts showed only the Mediterranean, whose coasts were de-

lineated with remarkable accuracy. Other features are exemplified in Plate 1.1, which shows an Italian work from the middle of the fifteenth century. Portolan charts typically showed deeply indented coastlines, symbolic sketches of cities, place-names at right angles to the coast, and a system of wind roses and intersecting lines that traced compass bearings for navigation. This type of map was very important in the earliest phase of European expansion across the Atlantic (see for instance Plate 2.1, which shows a conventional portolan chart view of Europe, now linked to a vision of the New World).

The portolan chart was a pragmatic document, drawn by using compass bearings, but without reference to any larger geographical scheme. It did not have to take into account the problems of representing the spherical surface of the earth on the flat surface of a map, because it covered a relatively small area. This problem of how to create an accurate and recognizable projection of the earth's surface had been summarized about 200 A.D. by Claudius Ptolemy. Ptolemy's geographical works were rediscovered in the early fifteenth century, circulated in manuscript, and then widely diffused at the end of that century in printed editions. For people imbued with the spirit of the Renaissance, or rebirth of classical antiquity, Ptolemy was to geography as Virgil was to poetry, or Aristotle to physics—that is to say, *the great authority*. Plate 1.2 shows Ptolemy's version of the world in a manuscript map drawn about 1460. This image has many weaknesses, some obvious, such as the great bulk of Africa, and some more subtle, such as the excessive length of the Mediterranean. But the Ptolemaic system also had the unique feature of demonstrating how to establish a network of coordinate numbers, one of latitude (north-south) and one of longitude (east-west), following which any location on the earth could be identified mathematically and plotted on a globe or map. This system was indispensable for the establishment of maps like the one shown on Plate 2.2, and indeed it is almost impossible to imagine how the world could have been cartographically knitted together without it.

Europeans of the sixteenth century thought of mapping in three distinctly different ways: *Geography* was the delineation of the earth at large, *chorography* the mapping of regions such as provinces, and *topography* the description of relatively confined spaces such as estates or towns. These distinctions are still helpful, and we shall follow them in going on to consider chorographical maps. These maps were relatively rare before 1500 but became abundant during the sixteenth century, particularly under the influence of kings and statesmen who wanted to "see" what their realms looked like. One of the most interesting series of chorographical maps was the set of English county maps drawn by Christopher Saxton in the 1570s, under the patronage of the English crown (Plate 1.3). These maps show the main towns and villages, with some attempt to delineate the topography and to pay attention to such fashionable innovations as gentlemen's hunting parks. Saxton's work eventually covered the whole of England and Wales, but since each of the counties was allowed a page of the

same size and since the counties differ considerably in extent, the maps were of widely differing scales.

Many other countries were mapped during this period using the province, or its equivalent, as the base. Then during the seventeenth century the Cassini dynasty in France put mapping at this scale on an altogether more accurate basis. The Cassinis created a new type of map by taking the country as a whole for their subject and by establishing a network of astronomically fixed observation points. Plate 1.4 is an example of the work of the Ordnance Survey, the English equivalent of the Cassini surveys. The county boundaries, which had formed the very basis of Saxton's survey, are now incidental to the plan of the work, and the whole country is covered at a large scale—that is to say, in great detail, with considerable accuracy. In North America, maps like those shown on Plates 3.2 and 3.3 were constructed using similar techniques, sometimes no doubt by topographers who had learned these techniques in the Old World.

At the topographical level, we have maps of particular localities. These had been drawn in various parts of Western Europe during the fifteenth century, often as part of legal proceedings. Such maps became more numerous during the sixteenth century and sometimes reached the sophistication of the plan shown in Plate 1.5. This was the period when artists like Albrecht Dürer and Leonardo da Vinci often toyed with the idea of producing landscapes for their own sake, no longer merely as backgrounds for portraits, and topographical skills emerged like the ones needed to show the forest in Plate 3.7. However, there are virtually no images of North America like the one shown on Plate 1.5, no doubt because the land was so abundant that it was not worth quarreling over to the degree of intensity implied in this fine drawing.

In England, actual surveying on the ground was carried out in very much the way shown on Plate 3.4, with the surveyor directing assistants who actually carried the sighting poles and chain, while the surveyor made notes that he could later plot into a sketch. Plate 3.4 also shows what an English-type metes and bounds survey looked like, running from one conspicuous feature, such as a quarry or prominent tree, to the next. Much of Western Europe had been laid out in a similarly irregular fashion, though in some places the rectangular Roman field layout still survived and in other places the land was still divided into the strips that were characteristic of farm plots in the lowlands of northern Europe. These strips had their counterpart in the New World in the *long lots* invented by French settlers (see Plates 3.6, 4.6, and 12.4).

Sometimes the surveyors worked their rough plats into quite elaborate estate-plans such as the one shown on Plate 3.5. This seems to have been essentially an English cartographic form, which emerged in the late sixteenth century and spread to Ireland, to the West Indies, and then to parts of North America. Such plans were commissioned by great (and often absentee) landowners, both to give them an idea of the layout of their estates and also to serve as an object of pride, for the plans were often very beautifully executed. Plate 1.6 shows a map of a similar genre from France. It delineates the thirty-fourth of one hundred *cantons,* or small administrative subunits, in a French parish. The plan is characteristically detailed, showing each individual house and field. In France, where the economic importance of forests was great both for hunting and for husbandry (acorns for pigs, wood for heating, chestnuts for eating, and so forth), there were many forest plans like the one shown on Plate 1.7. Interestingly enough, these seem to have spread neither to England nor to the New World; certainly in the latter area the forests were far too extensive to require close management and scrutiny.

Towns presented special problems to the European cartographer, for then as now they had an extraordinary amount of informa-tion to be mapped. Three types of plan were eventually developed in order to depict towns:

1. The *profile,* or view of a town, as seen from the ground as one approached it.
2. The *bird's-eye view,* in which the town was seen obliquely from an aerial vantage point.
3. The *planimetric plan,* in which the town was shown directly from overhead.

All these types may be seen among the plates of the six-volume *Civitates Orbis Terrarum,* which contains images of most of the world known to western Europeans, collected by Georg Braun and Franz Hogenberg between 1572 and 1617. As time went by, the planimetric view was used most often, because it is the most comprehensive and accurate. But profiles and bird's-eye views continued to be used and indeed made their way to the New World. As Plate 1.8 shows, the profile could give a very good impression both of a city and of the way it lay on the land. Of course, it could not show the precise line of the streets, most of which remained hidden behind walls and buildings. A well-conceived bird's-eye view, like the one on Plate 1.9, in some ways combined the best features of the other two forms. It could be interpreted without any training on the part of the observer, and yet it gave a good impression of the actual internal shape of the urban area. Plate 1.10 shows a relatively early plan, in which the streets, rivers, and fields are drawn as if from directly above, while the buildings have a variety of different perspectives. As we shall see in Chapters 3, 6, and 7, all these types of city views had progeny in the New World.

There is, finally, one type of map that is often neglected—the nautical chart. As we have seen, the portolan chart was the primary maritime cartographical form inherited from the Middle Ages, but as time went by, the techniques developed on land were also applied to developing a new generation of sea charts. The Dutch were the first to do this, in the late sixteenth century. In the seventeenth century they were joined by the English and French, who began making charts of hitherto unknown accuracy, often known under the generic title of *admiralty charts*. Plate 1.11 shows one of these, illustrating well how such charts could contain a great deal of information about features on land as well as at sea.

When they came to the New World, then, the various European peoples were well equipped to map its complex shapes. However, only about half the chapters of this book deal with map forms inherited from the Old World. Many of the map types discussed here were more or less New World inventions, like the township and range system (Chapter 4), county maps and atlases (Chapter 8), railroad and fire insurance maps (Chapter 9), and highway maps (Chapter 11). As always, the cartographic form faithfully reflected the developments of history at large; whereas at the beginning of the period the New World had been in tutelage to the Old, by the end of it she was inventing new forms and exporting them back to Europe.

SOURCES DEALING WITH THE USE OF EUROPEAN MAPS AS HISTORICAL DOCUMENTS

Andrews, J. H. *History in the Ordnance Map*. Dublin: 1974. Analysis of the uses of this British map type for the historian.

Baker, A. H. R. "Local History in Early Estate Maps." *Amateur Historian* V (1962).

Carr, A. P. "Cartographic Record and Historical Accuracy." *Geography* XLVII, pp. 135–46.

Coppock, J. T. "Maps as Sources for the Study of Land Use in the Past." *Imago Mundi* XXII (1968), pp. 37–49. Survey of eighteenth- and nineteenth-century maps with helpful plates.

Grosjean, G. "Landkarten als Kultur- und Geistesgeschichtlichen Dokumente." *Schweizerisches Gutenbergmuseum* LIV (1968), pp. 47–177. Reflections by one of Switzerland's leading historians of cartography.

Harley, J. B. *Maps for the Local Historian: A Guide to the British Sources*. London: 1972. This book tries, in 86 pages, to do for British history what the present manual is attempting to do for U.S. history.

———. "Ancient Maps Waiting to Be Read." *The Geographical Magazine,* 1980. Some examples of maps as sources.

Hayes McCoy, G. A. "Contemporary Maps as an Aid to Irish History." *Imago Mundi* XIX (1965), pp. 32–37. Description of some newly discovered manuscript maps, together with general reflections on maps and Irish history.

Kain, R. C. P. "Compiling an Atlas of Agriculture in England and Wales from the Tithe Surveys." *The Geographical Journal* CXLV (1979), pp. 225–41. Description of the historical uses of the nineteenth-century tithe surveys, a prime source.

Koeman, C. "Levels of Historical Evidence in Early Maps." *Imago Mundi* XXII (1968), pp. 75–80. An attempt to grade maps into six categories of historical reliability.

Lobel, M. D. "The Value of Early Maps as Evidence for the Topography of English Towns." *Imago Mundi* XXII (1968), pp. 50–61. Short but informative survey of the material by a scholar who has worked extensively in the history of urban development.

Wallis, H. "Early Maps as Historical and Scientific Documents." In *Nordenskiöld Seminar,* ed. K. Hakulinen and A. Peltonen. Helsinki: 1981.

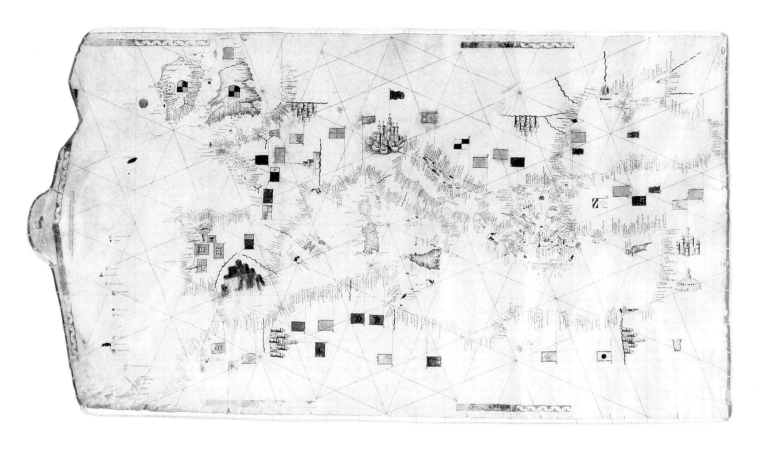

Plate 1.1. Portolan Chart of the Mediterranean area by Petrus Roselli (1456).

Plate 1.1a. Modern outline of Europe and the Mediterranean.

A Medieval Chart of the Mediterranean

This type of map first appeared about 1200. Its main features changed remarkably little down the centuries until about 1600, when it faded out. It is drawn on animal skin, with the neck of the animal to the left, and shows the Mediterranean with great fidelity. Scholars differ about how this accuracy was achieved—and a chart of 1200 would be just as accurate as this one of 1456—but it seems likely that the use of the magnetic compass accounts for it; certainly the first appearance of the portolan chart and of the magnetic compass in the Mediterranean are more or less simultaneous.

The map is for sailors, and consequently shows little or no interior detail. Ports, however, are marked in great detail, at right angles to the coast, and a few great cities are shown in profile. Here, Venice is particularly prominent. A network of lines covers the map, and identifying flags fly above the territories of the greater rulers. There are no indications of latitude or of longitude, for this was a map form innocent of astronomical readings. Note that cartographic knowledge tends to fade out in the North Sea and around the top of the British Isles, though the Atlantic islands are shown in some profusion. This progressive charting of the Atlantic islands in fact pointed towards the eventual mapping of what lay beyond them. It was with this cartographic genre inherited from the Middle Ages that the Europeans would first map their New World.

Source: Michel Mollat du Jourdin and Monique de La Roncière, Sea Charts of the Early Explorers *(New York: 1984).*

Plate 1.2. World Map according to Ptolemy from the Ebner Manuscript. Reproduced from the *Geography of Claudius Ptolemy,* ed. E. L. Stevenson, 1932. Rare Books and Manuscripts Division, The New York Public Library, Astor, Lenox and Tilden Foundations.

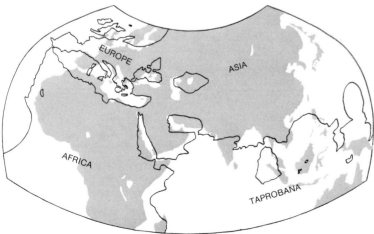

Plate 1.2a. Ptolemy map superimposed on a modern map of the same area.

The World According to Claudius Ptolemy

Claudius Ptolemy, who lived in Alexandria about 200 A.D., summarized the geographical and cartographical learning of Greek and Roman antiquity in his *Geography*. Most of this work consisted of latitude and longitude references for a great variety of places, but it was also possible, using these references, to construct maps according to Ptolemy's instructions.

With the collapse of the ancient world, Ptolemy's cartographic information was lost to Europe, but it survived in various manuscripts, some of which began to be rediscovered in the fifteenth century, as part of the drive to recover the works of classical antiquity known as the Renaissance. The world map shown here comes from the Ebner Manuscript of about 1460, now in the New York Public Library; it is named after Wilhelm Ebner, who owned it in the eighteenth century, and gives a typically Ptolemaic version of the known world.

Notice that the Mediterranean is quite well visualized, though not as well as on the portolan charts. The area to the east is also rendered faithfully, but Ptolemy's knowledge tends to break down east of the Persian Gulf. Here the great island of Taprobana incongruously floats in the Indian Ocean, perhaps representing a conflation of information about Ceylon and Sumatra. The Malay Peninsula is sketched in, and Africa is left as a solid block.

One element is of great importance, and that is the figures for latitude and longitude along the edges of the map. Even if Ptolemy's own knowledge of the world was limited, he provided the mathematical basis upon which subsequent cartographers could build (see for instance Plate 2.2). He was also crucial in forming the geographical consciousness of that generation of Europeans who directed the first wave of exploration to the New World, for his *Geography* appeared in printed form with maps in 1477, and thereafter went through many editions in many different European languages.

Sources: Leo Bagrow, History of Cartography, *2nd ed. (Chicago: 1985); and Geography of Claudius Ptolemy, trans. and ed. E. L. Stevenson (New York: 1932).*

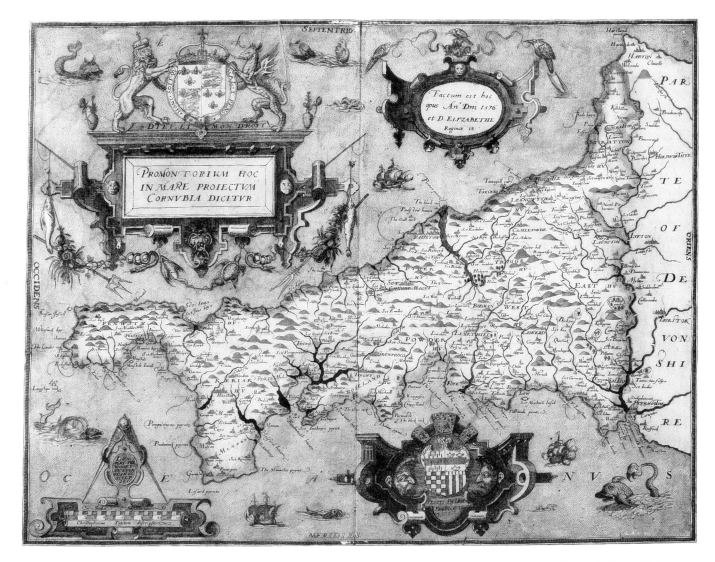

Plate 1.3. Cornwall, from Christopher Saxton's maps of the counties of England and Wales (London, 1579).

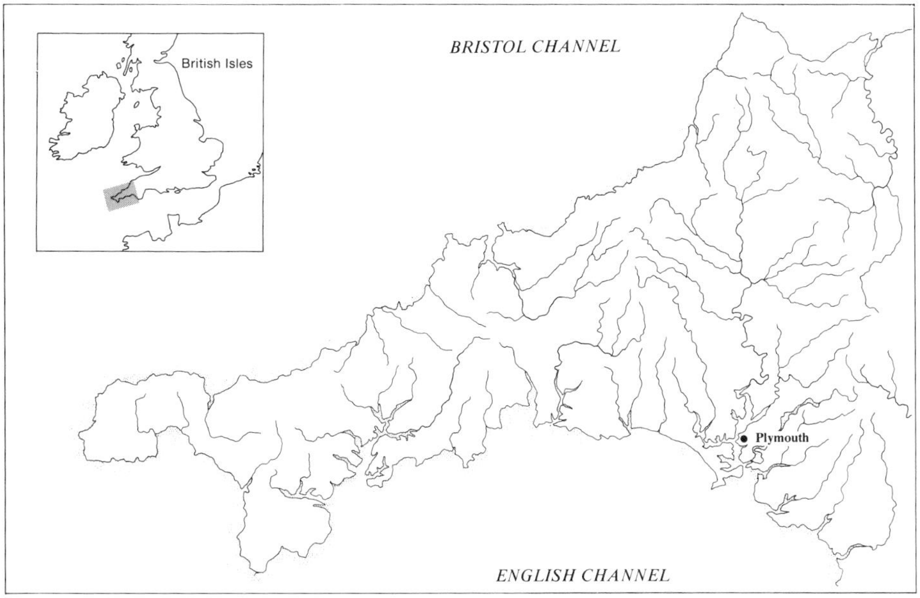

Plate 1.3a. Modern map of Cornwall.

BRISTOL CHANNEL

British Isles

Plymouth

ENGLISH CHANNEL

An Example of a Chorographical Map
from England, 1576

This is a fine example of a sixteenth-century provincial, or chorographical, map. It was compiled by Christopher Saxton, an estate surveyor, working under the protection of the English Crown, and is one of a series covering the whole of England and Wales. These maps were published in 1579, and long formed the base for subsequent maps of the area.

The cartouche in the upper left gives the map title in Latin ("This Promontory thrust out into the sea is called Cornwall") surmounted by the royal arms. On the upper right are details about the data of compilation; lower right are the arms of Saxton's patron, Thomas Seckford, and lower left is a scale. All Saxton's maps were oriented to the north, perhaps following the example of Ptolemy and of the portolan charts. Many other provincial maps of this period, however, continued to be oriented however the cartographer thought fit; generally in the direction looking "out," so to speak, from wherever a great city happened to be, or even in the direction that would best fit the page.

This map is fairly accurate in its general outline, as well as in the course of the rivers, and may well have been constructed by drawing sight lines from the tops of prominent places such as church towers or beacon hills. It shows the rivers in generalized form, and the hills rather grossly, for there was as yet no system of contour lines to show land elevation. The size of the lettering indicates the relative importance of features; thus "Plymouth" (bottom right) receives large lettering as befits a large town.

It is easy to see the shortcomings of this rather approximate map, and to imagine how much more useful would be the ones drawn in the seventeenth century. But as a first approximation it is highly satisfactory, and all over Europe, in the later sixteenth century, people began to locate themselves spatially by means of maps like this, learning skills that were taken to the New World in maps like those shown on Plates 2.4, 2.7, and 2.8.

Source: Sarah Tyacke and John Huddy, Christopher Saxton and Tudor Map-making *(London: 1980).*

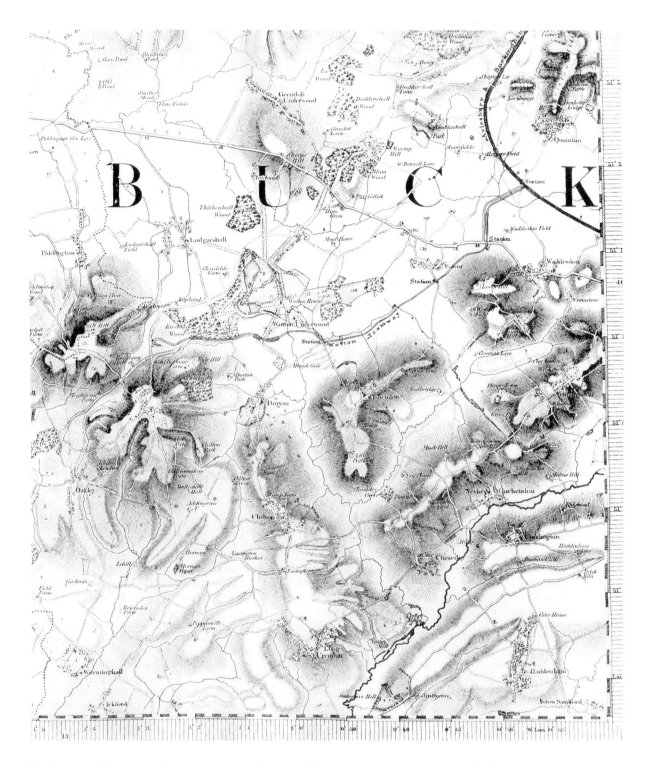

Plate 1.4. Detail from sheet 61 of the late-nineteenth-century Ordnance Survey map of England and Wales.

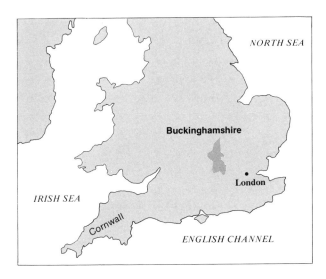

Plate 1.4a. Location map showing Buckinghamshire and Cornwall.

An Example of a Large-Scale Cassini-Type Map

Leaving provincial maps, we come down to the more detailed surveys that began to be drawn in the seventeenth century, originating with the large-scale survey conducted by the Cassini dynasty, under royal patronage, in France from 1670 onward. Almost every European country was eventually covered by maps of this kind, which, disregarding the former provincial boundaries, simply took the country as a whole and overlaid it with a checkerboard of interlocking maps.

In England this development came relatively late, so well had Saxton worked, and it was not until 1791 that a survey directed by the (Army) Board of Ordnance began what would become known as the *Ordnance Survey*. This eventually covered the British Isles at scales of as large as one inch to the mile, giving details not only of towns and hamlets but even of individual farms.

Our plate shows a detail from sheet 61 (called "Banbury") from the Ordnance survey map issued about 1890, though virtually unchanged in appearance for about a century. Across the top run some letters from the name of the county (BUCKinghamshire), and round the edge of the whole sheet (not shown here) are the figures for latitude and longitude.

The system for indicating heights is still rather rudimentary, consisting of a sort of representational shading called *hachuring,* but the level of detail is impressive. Villages like Ludgarshall and Wotton Underwood have all their houses shown, as well as copses, isolated farms, roads, and alleys of trees. Across the middle of the map a modern feature intrudes, the Wotton Tramway, no doubt a form of railroad. Maps of this kind could be used for a great variety of purposes, from planning new roads to moving large bodies of soldiers, and they became indispensable to European governments. Their counterpart in North America was no doubt the survey associated with the township and range system discussed in Chapter 4, and the USGS quads, discussed in Chapter 10.

Sources: Marc Duranthon, La Carte de France *(Paris: 1978); and J. B. Harley,* Ordnance Survey Maps: A Descriptive Manual *(Southampton: 1975).*

Plate 1.5. Drawing of the mid-sixteenth century showing Wotton Underwood in the Buckinghamshire countryside. The Huntington Library, San Marino, California.

An Image of the English Tudor Countryside

This drawing was made about 1565, in the course of a dispute between the inhabitants of Wotton Underwood (near village) and Ludgarshall (far village) about the common land lying between them. It offers us a rare glimpse of the Tudor countryside, with its dense woods and relatively few enclosed fields, and nothing much in the way of roads. In the middle on the left is a pond, and on the right a windmill of the characteristic *post* type. Each village has its church, surmounted with a cross, and at the left-hand end of Wotton Underwood is the house of the Grenvilles, squires of the village. As yet their residence is quite modest, hardly to be distinguished from those of the other villagers.

It is instructive to compare this view with the map reproduced as Plate 1.4. Here we visualize the same area from roughly the same angle, and can see what has happened to the disputed common: It has been subsumed into the park of Wotton House, providing a sort of cartographic symbol for the emergence of the great landowner in the English countryside. This drawing is of interest not only as an historical document, but also as an early cartographic type. It was in the sixteenth century that the skills required to make accurate topographical drawings, which perhaps developed out of the work done by the miniaturists of late medieval Europe, became relatively commonplace, particularly in Italy, Germany, and England. Eventually they were carried to the New World, where, however, they were applied to towns (Chapter 6) rather than to countrysides.

Source: "A Shakespeare Haunt in Bucks," Shakespeare Quarterly *V (1954), pp 177–78.*

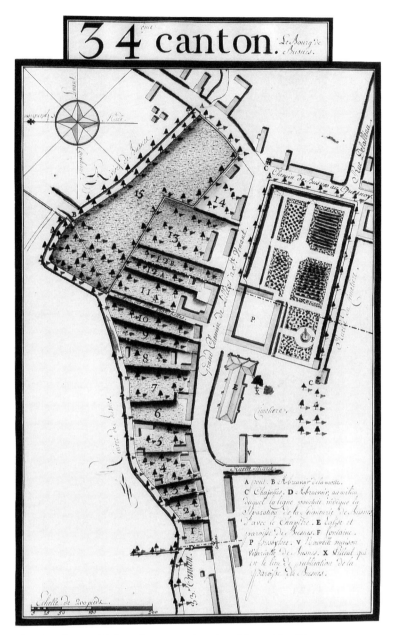

A Large-Scale Topographical Map of the Countryside, 1782

This is a page from an atlas showing the parish of Busnes, in northern France, drawn in 1782 by Christophe Verlet, a surveyor. Such atlases, which are rare, are the French counterpart to the English estate plan, of which a perfect New World example is shown as Plate 3.5. The parish of Busnes was divided into 100 *cantons*, or administrative subdivisions, and each canton had its map, at a very large scale.

Our plate shows canton 34, which contained the church (large building marked "E") and priest's house ("P"). The roads are shown, and so are the smallholdings at the back of each house; the little river ("Riviere de Busnes") winds along the left-hand edge, from top to bottom. A map like this exposes the most intimate spatial arrangements of the eighteenth-century village. We can imagine the peasants emerging from their houses on the high road ("grand chemin") and strolling across to hear Mass in the church. After the Mass they might well linger under the large linden tree marked "X"; here, as the legend tells us, was where the king's edicts were read ("lieu de publication de la paroisse de Busnes"). It would be a fascinating exercise, to walk the ground of present-day France with a map like this in hand.

Source: P. D. A. Harvey, The History of Topographical Maps *(London: 1980).*

Plate 1.6. Canton 34 from the manuscript atlas of the parish of Busnes, near Lille in northern France, by Christophe Verlet (1782).

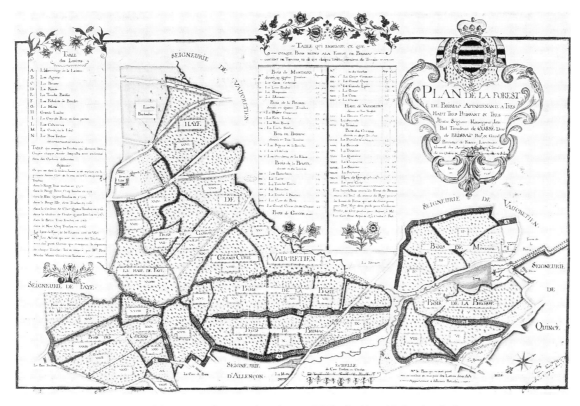

Plate 1.7. Manuscript plan of the Forest of Brissac, by Vincent (1756). Archives Nationales, Paris.

An Eighteenth-Century Forest Plan

Sherlock Holmes once observed that in detective work, it is sometimes important to listen to the dog that does not bark. So it is with the comparative history of cartography. Whereas in France plans showing agrarian estates were relatively few and relatively late, seeming to show an agriculture firmly wedded to traditional practices, there were on the contrary many plans of forests, as there were in Germany (though few in England and fewer in the Netherlands).

The plan shown here delineates the forest of Brissac, belonging to the Duke of Cossé-Brissac, and was drawn in 1756. The forest is divided into various "bois," or woods, and the central table lists the size of each. The table on

the far left contains a list of "lisieres" or forest paths ("rides"), with a key to the colors indicating when work needs to be done on various parts of the forest.

This is therefore a working tool for keeping track of what needed to be done in order to get the most out of the great forest. Maps of this kind were also common in Germany, but they did not cross the Atlantic. For there, the immense forests were seen either as a mere impediment to agriculture, or sometimes as a boundless reservoir for game; in either case, they did not merit the close attention that the Duke of Cossé-Brissac paid to his domain.

Source: Michel Devèze, La Vie de la Forêt Française au XVIe Siècle *(Paris: 1961).*

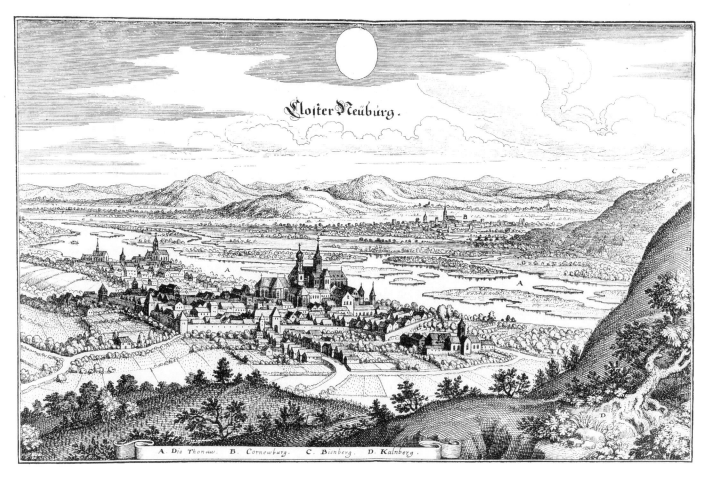

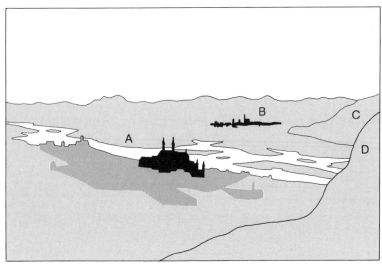

Plate 1.8. "Kloster Neuburg," from Matthaeus Merian, *Topographia Provinciarum Austriacarum* (Frankfurt, 1649).

Plate 1.8a. Diagram of Plate 1.8.

A Seventeenth-Century Town Profile

This is an example of a profile or town view taken from the ground approaching the place. It forms part of the immense collection of town views published in the seventeenth century by Matthaeus Merian at Frankfurt, and is typical of them in the delicacy and elegance of the engraver's touch. A view like that gives us a very good idea of the way in which a town fitted into its surroundings. Kloster Neuburg, in the foreground, lies just outside Vienna, in Austria. Across the middle distance flows the Danube ("A" on the key), with the little town of "Cornewburg" ("B") on the other side of it, and the two hills, "Bisnberg" and "Kalnberg" ("C" and "D"), rising on the right.

We have a good idea of the extent of the town, which is spilling out of its medieval walls towards us into what will one day be called *suburbs*. We also get a good notion of the disposition of the main monuments in the town, which indeed is dominated by its great church, no doubt belonging to the abbey ("Kloster") around which the town developed. What a profile like this does not show is the internal disposition of the streets. We can make a rough guess at it from the position of the houses, but we should need a more vertical vision fully to comprehend the layout inside the walls.

Source: James Elliott, The City in Maps: Urban Mapping to 1900 *(London: 1987).*

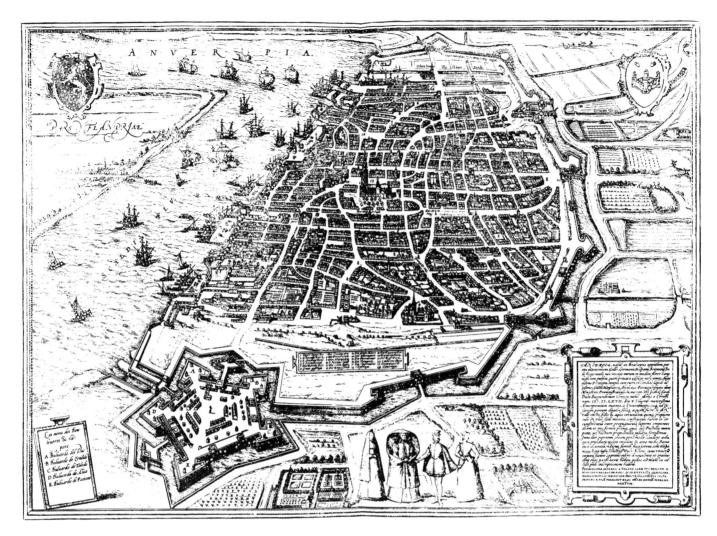

Plate 1.9. Antwerp from the *Civitates Orbis Terrarum* of Georg Braun and Franz Hogenberg (Cologne, 1572–1617).

Plate 1.9a. Location map showing Antwerp and the River Scheldt.

A Sixteenth-Century Example of a Bird's-Eye View

The 1570s saw the beginning of two great and similar publishing ventures in the map world. Abraham Ortelius brought out the first edition of his *Theatrum Orbis Terrarum* (Antwerp, 1570), and Georg Braun and Franz Hogenberg published the first volume of their *Civitates Orbis Terrarum* (Cologne, 1572). Whereas the *Theatrum* offered a well-conceived collection of maps of different parts of the world in standard format, the *Civitates* brought together, also in standard format, a rather heterogeneous collection of city views. However, each fulfilled a great need and was a great commercial success. The *Civitates* finally ran to six volumes, and its derivatives continued to be published far into the seventeenth century, for it became almost a matter of course for every town of any size to have its place in this great collection.

We could find any of the three types of town plans in this extensive publication, but have chosen a bird's-eye view of the great city of Antwerp. The artist has adopted an angle that allows him to show the recent citadel very prominently, at the bottom left, and has also listed the bastions in the newly constructed wall. Away to the left, ships crowd the Scheldt River and the wharves, bringing goods from all over Europe to this great commercial and industrial center, which by now also handled spices from the East. The advantage of the bird's-eye view over the profile shown in Plate 1.8 is evident; here we not only seize on the main monuments like the citadel and cathedral (center), but can also see the layout of the streets. Indeed, it is possible to this day to navigate in central Antwerp using nothing but this marvelous plan.

Sources: J. J. Murray, Antwerp in the Age of Plantin and Breughel *(Norman, Okla.: 1970); and R. A. Skelton, ed.,* Braun and Hogenberg's Civitates Orbis Terrarum 1572–1618, *3 vols. (Amsterdam: c. 1980).*

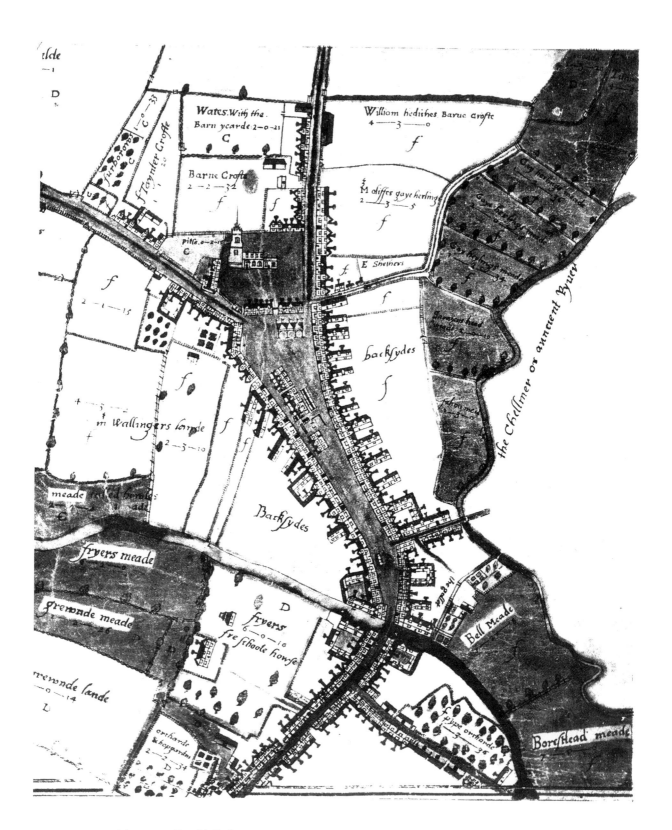

Wates. With the.
Barn yearde 2—0—21
C

Barne Crofte
2—2—32
f

William hediſhes Barue Crofte
4———3———0
f

M cliffes gaye herlings
2———3———5
f

Furforowle
c

S. Faynter Crofte
f

pitle. 0—2—15
C

E Shethers
f

f

backſydes
f

Barkſydes

m Wallingers lande
2—3—10

meade ſloid bottes

fryers meade

grewnde meade

grewnde lande
0—14
D

fryers
6—0—10
D

freſchoole howse

orcharde
ſchoppardine
2—1—34
D

Bell Meade

BoreſHead meade

the Chellmer or auncient Bruet

A Sixteenth-Century Example of a Planimetric Map

Plate 1.10. Detail from a manuscript map of the Manor of Bishop's Hall by John Walker, 1591 (D/DM P1). Reproduced by courtesy of the Essex Record Office, Chelmsford, England.

Here is an example of an early planimetric view of a town, Chelmsford, in eastern England (see Plate 1.4a for location of Chelmsford). The main road, rising in a great fork to the top of the image, and the river, curving across the bottom half of it, are drawn as if seen vertically from above, and at the top there is a scale (impossible with profiles or bird's-eye views) for accurate checking of the various parts. The various fields all have their distinctive names— "Fryer's Meade," "Barne Croft," "Backsydes," "Grevonde Lande," and so forth—describing not only their ownership but also their type. The "meades" by the river are meadows, the croft rough grazing, and so forth.

The planimetric image is not one "natural" to the human brain, and so it was often difficult to maintain it throughout early plans. Here it breaks down for the church, seen in profile, and indeed for the houses lining the streets, seen in a variety of profiles. Still, we can see what the surveyor was trying to do, and can acknowledge that he created a plan that is not only a fascinating evocation of a Tudor town (little changed, incidentally, in its main lines to the present day), but also a map type that would have many successors in the New World (see particularly Chapter 7).

Sources: A. C. Edwards and K. C. Newton, The Walkers of Hanningfield *(London: 1985); and F. G. Emmison, ed.,* Catalogue of Maps in the Essex Record Office 1566–1855 *(Chelmsford: 1947).*

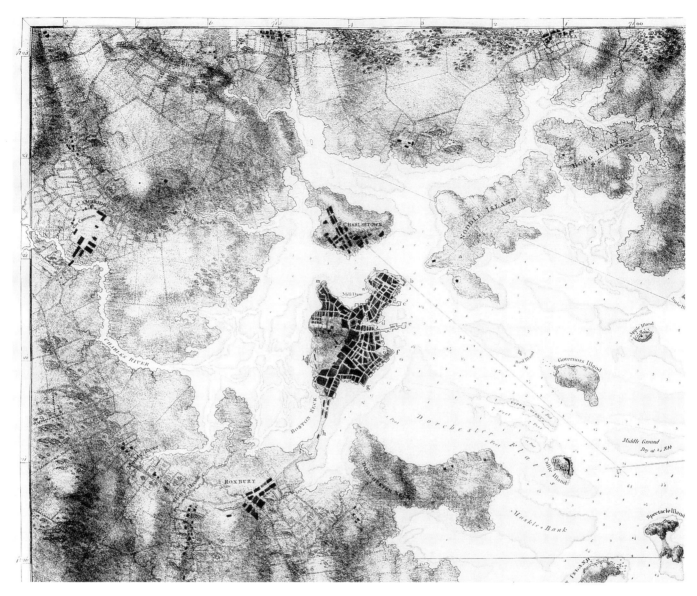

Plate 1.11. Detail from Joseph Des Barres's map of Boston Harbor, in *The Atlantic Neptune* (6 vols., London, 1774–84).

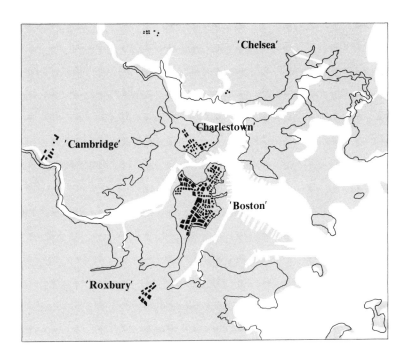

Plate 1.11a. Plate 1.11 superimposed on modern outline of Boston.

An Eighteenth-Century Nautical Chart

On Plate 1.1 we saw an example of the type of nautical chart with which the Europeans entered upon their great period of expansion. Towards the end of the sixteenth century the Dutch began to produce nautical atlases that were the counterparts of the city plans of Braun and Hogenberg, and of the general maps of Ortelius. But it was not until the late seventeenth century that a recognizable new nautical style emerged, perhaps as the counterpart to the new Cassini-type land maps (Plate 1.4).

These so-called *Admiralty maps* were commissioned by the great naval powers of northern Europe: the Dutch, the French, and the English. They are characterized by a high level of accuracy, not only in their information about nautical matters such as anchorages, soundings, and currents, but also in the details that they sometimes offer of adjacent land areas.

Our plate shows Boston Harbor as drawn by Joseph Des Barres in *The Atlantic Neptune: Published for the Uses of the Royal Navy of Great Britain,* 6 vols. (London: 1774–1784). Notice that this is not only a town plan of the cities of Boston and Charlestown, but also a fairly detailed indication of the layout of the surrounding countryside, from Roxbury at the bottom of the detail, to Cambridge on the left of it, and Chelsea at the top right. In North America, great quantities of maps of this kind were produced by the U.S. Hydrographic Office, and they are often interesting to historians as well as to sailors (see reference in Chapter 9).

Source: Derek Howse and Michael Sanderson, The Sea Chart *(New York: 1973).*

CHAPTER TWO

Maps of the Age of European Exploration

David B. Quinn

The first maps of North America drawn by Europeans are difficult to read and often misleading. One must look at them critically to extract their content and meaning, for these maps, like their European forebears, were often a mix of real and imaginary features. Many are also initially confusing because of their strange orientation and their distortion of landmasses that are now familiar to us. Even with their most obdurate peculiarities however, these maps, when compared to modern atlases and detailed sheet maps, reveal explorers' knowledge of and ideas about the New World.

North America was unknown to European cartographers before the last decade of the fifteenth century, even though there was steady progress in mapping the Atlantic islands up to that time. The maps and charts made before the actual discovery should not be ignored, for they illustrate some of the problems that had to be faced once discovery began to take place; what, for instance, was the relationship between the Old World and the newly found territories, and how could this globe-girdling image best be expressed on a flat surface like a map? After 1492, the discoveries were partial and the cartographic representations of them imperfect. As mapmakers gathered the threads of new information into maps, the shape of North America changed with increasing rapidity. How extensive was the new land, how wide, how long? These were questions that cartographers tried to answer in very different ways, so that for a long time some thought that a great ocean led to the west out of what would be the Carolinas, and others long believed that California was an island. The problems of giving a truly accurate impression of the interior were by no means solved by 1700; indeed, in some respects they were only beginning to be addressed, and we must look to the later chapters of this book for an account of the progress that was eventually made.

The earliest sea charts and most of the early maps depicted what explorers knew or guessed about North America and the Atlantic Ocean. Navigators in the fifteenth and much of the sixteenth centuries took no account of the curvature of the earth, which indeed had not been of much relevance in the Mediterranean, where many of them had made their initial voyages. Thus their maps were constructed on a system that could not possibly represent large areas such as North America with any approach to verisimilitude. However, maps of small areas could be and were made accurately while ignoring curvature. The problems of latitude and longitude were more difficult to overcome. By the time parts of North America began to appear on the map, latitude could be determined by observation of sun and stars, but taking sights from the sea, with its inherent instability, could produce substantial inaccuracies. At sea accuracy might approach only within two degrees (140 land miles), while close observation on land might be accurate to within twenty minutes of arc or even less (fifteen to twenty miles, perhaps). Thus general maps or charts created by sailing along the coast were often inaccurate in representing north-south relationships, although detailed maps made on land could be reasonably accurate. Longitude was more difficult still. It could not be determined by instruments at sea until about 1750. Rough-and-ready dead reckoning could give an approximation, but only a period of observation of sun and stars on land could produce a map approaching reality.

One of the major problems in examining charts of the Atlantic Ocean from the fourteenth to the seventeenth centuries, and in some cases even later, is that imaginary and real features were depicted side by side, and it is extremely difficult to distinguish genuine from false or mistaken representations on these early charts. The east coast of mainland North America from the tip of Florida to north of Labrador and into the Davis Strait was explored and mapped by 1600, but few examples of the coastline agree with each other. Degrees of accuracy and inaccuracy have to be carefully worked out by comparison of one map with another. Much progress was indeed made in the seventeenth century as ships and settlers used the shores of eastern North America for trade and settlement, so that by 1700 the picture of the east coast is reasonably clear and accurate.

The west coast is very different. Spanish expeditions defined the west coast as far north as about 44 degrees (about the latitude of

Eugene, Oregon) between 1540 and 1603, but no new information was charted during the following century, so that conjecture took the place of knowledge. Indeed, during the seventeenth century the vast "island" of California made its appearance, surviving until after 1700 on some maps. If the west coast was known only in the grossest outline, much of the interior remained entirely unmapped until the nineteenth century. Knowledge about the northern half of the new continent came slowly, as explorers and eventually settlers nibbled away at the edges of this vast area.

The first region to be mapped on the mainland was Florida, into which the Spaniards expanded from their first bases in the West Indies, so well shown on the Juan de la Cosa map of 1500 (Plate 2.1). Already on that map we read about the discoveries of John Cabot around Newfoundland in 1497 and 1498; knowledge of the intervening coastline came from a series of voyages, of which the one by Estéban Gomez in 1525 is the best known. Much of the knowledge was plotted in cartographic form by German mapmakers, one of whom, Martin Waldseemüller, first set out a map affirming unambiguously the existence of a new continent (Plate 2.2).

We can follow the progress of Spanish exploration in the early sixteenth century through the records of a very remarkable institution established at Seville in 1508 by the Spanish monarchy. This was the navigation school attached at that time to the House of Trade *(Casa de la Contratación),* the body that regulated all relations with the newly found territories. This school had skilled cartographers and pilots, and Spanish captains were obliged to report to them upon leaving Seville for voyages and upon their return. The cartographers of the Casa de la Contratación then used this information to draw up a master map, or *padrón real,* on which successive discoveries were recorded. Plate 2.3 shows a detail from the padrón real as it stood about 1529, revealing the extent of Spanish knowledge at that time about Florida and the east coast. Along the coast are four "tieras," or lands. In the north is the "Tiera de Labrador," discovered, as the legend says, by the English. To its south lies the land of Corte Real, discovered by the Portuguese brothers of that name; south again are the lands of Estéban Gomez and of Lucas Vasquez de Ayllón, on voyages for Spain.

The French produced their own profile of the east coast after 1524, following the voyage of Giovanni da Verrazzano in the *Dauphine*. Verrazzano's observations in fact gave rise to a major error in North American cartography, when, looking across the Outer Banks chain of islands from the Atlantic Ocean, he mistook Pamlico Sound for the Pacific Ocean. For some years various maps reflected this mistake, showing the Pacific Ocean as extending right across where North America in fact lies. In other ways, though, Verrazzano's voyage was very fruitful; he became the first European to enter and describe New York Bay and the Hudson River, and eventually to reach Newfoundland. Between 1534 and 1543 the French also penetrated the Saint Lawrence Valley through the voyages of Jacques Cartier and Jean de La Roque, sieur de Roberval; these discoveries were reflected on the marvelously decorative maps produced during the 1540s and 1550s by the so-called *Dieppe School*. This was a group of cartographers, working in Dieppe on the Channel coast, who, under the initial guidance of the Portuguese, produced large and beautiful manuscript world-maps, showing the latest progress of French and other explorers.

The Gulf of Mexico is another area whose general outline was established early, in this case by the expedition led by Alonso Alvarez de Pineda in 1519. The resulting *Pineda chart* correctly shows the main outlines of the Gulf, its relationship to Cuba, and the shape of the Yucatan Peninsula. Here, as in some other cases, the earliest cartographers worked better than their successors, for on subsequent maps the Yucatan Peninsula was often shown as an island. The area inland from the Gulf was in-vestigated by the expedition led by Hernando de Soto between 1539 and 1543. Members of the expedition were probably interrogated upon their return by the royal cartographer, Alonso de Santa Cruz, and the map that he then drew offers us the first detailed, if imperfect, image of the rivers and Indian settlements of much of the vast area that drains into the Gulf from the north.

Toward the end of the sixteenth century, the English began settling and mapping the east coast of North America. During 1585–1586 the coast and hinterland of what would become Virginia, between Cape Lookout and Cape Henry, was mapped by John White and Thomas Harriot (Plate 2.4). White and Harriot were an exceptional team. White had served with the explorer Martin Frobisher as a figure and topographical painter, and Harriot had much experience both as a mathematician and as a surveyor by land. They probably used a line, a compass, a plane table, and some kind of sighting instrument, and worked around the coast and inlets by boat. Their map is the first to have been made by Europeans in North America following a careful field survey, and it is exceptionally accurate, as our comparative map shows (Plate 2.4a).

Almost at the end of the century a great world map was published in England, probably by Edward Wright, summarizing knowledge of North America at that time (Plate 2.5). His version of the northern regions took advantage of information from the voyages of British navigators Martin Frobisher (1576–1578) and John Davis (1585–1587); his delineation of Newfoundland and the Saint Lawrence is very good, and he inserts a lake to the west of the Saint Lawrence, no doubt following French cartographic tradition of the 1580s. He foreshortens, in Spanish style, the profile between the Penobscot and the Roanoke area newly surveyed by John White, but he draws the coast clearly and links it sensibly with Florida. On the west coast he shows the peninsula of California and a substantial part of the coast of upper California. He does not attempt to fill in gaps to the northwest or north, but North America is given something like its true width. Wright was fully aware of the older Spanish sources, of French explorations, and of the most recent English discoveries, and he combined this information into a map that shows North America as accurately as it could then be drawn.

The early years of the seventeenth century brought rapid developments in the definition of both the coastline and the interior of eastern North America. Between 1604 and 1607 the French explorer Samuel de Champlain investigated the coast between southern New England and the head of the Bay of Fundy. He made sketches of many harbor entries and compiled a general map of his discoveries, presumably in 1607–1608; the manuscript of this map, which was not published at the time, is now in the Library of Congress. This map was far superior in accuracy and execution to any extant coastal map of any part of North America.

Almost at the same time, Captain John Smith was surveying the Chesapeake Bay and the numerous important rivers that flow into it. His survey techniques were not as sophisticated as those of Harriot and White, but he was determined to outdo them in the detail that he accumulated. Versions of his map were circulating in manuscript from 1608 onwards, and it was published in 1612 (Plate 2.7). This map is particularly remarkable for the number of Indian village sites that were included with some precision, various ones being indicated as tribal capitals.

Champlain returned to the Saint Lawrence River in 1608, and his stay there until 1611 enabled him to produce an exceptionally valuable map of the Saint Lawrence River basin, published in 1613. He paid much attention to the charting of the island of Newfoundland, the Gulf of Saint Lawrence, the river, and its tributaries, so that the achievements of the Cartier-Roberval voyages, extending inland to the rapids beyond Montreal, were revised and brought firmly under control; this was a semi-

nal map, bringing together much previous work in a coherent whole.

Some of these early seventeenth-century ventures led to significant penetration of the back country. We do not have the English explorer Henry Hudson's chart of the Hudson River from 1609, but it appeared in a remarkable manuscript map, made in England before February 1611, and sent to Spain in that month. Known from the name of the ambassador who sent it as the "Velasco map" (Plate 2.6), it represented an English conflation of the work of Smith, Hudson, and Champlain before any of their work had been published. Where the information came from has remained something of a mystery, but it was clearly the result of successful espionage by the English, and its capture by the Spaniards was a diplomatic coup. It remained unknown until the twentieth century, but has now been reproduced several times, in whole and in part. If some parts of the coast, the Eastern Shore, and the Cape May–Hudson River and Hudson River–Buzzards Bay sections were still imperfect, and the location of Lakes Champlain and Ontario based on guesswork, its achievement was nonetheless outstanding, for it tied together information that until then had been available only in scattered form.

The situation in the seventeenth century was very different from that of the sixteenth. Already works such as Smith's *Map of Virginia* (1612) and Champlain's *Voyages* (1613) were almost as much concerned with propaganda for colonies as with the representation of the North American continent; they were the almost indispensable preliminaries to settlement. John Smith's map of New England in his *Description of New England* (1616) was of the same character. Just as Verrazzano in 1524 had put the idea of *New France* onto the map, so too the invention by Smith of the name *New England* opened up that area to English colonization.

The French worked steadily on. Champlain elaborated his own map of 1613 with material from further explorations, from English discoveries, and from Indian tales of the Great Lakes in the final 1632 version of his *Voyages*. Other maps of New France were made after the French recovered it from England, during the time when priests and trappers were slowly penetrating the country around the Great Lakes. Nicolas Sanson summarized much of this progress by publishing in 1656 a map of New France that extended to Lake Superior, which preceded by only seven years the assumption of French royal authority over Canada. On the English side Newfoundland attracted cartographic attention when it, too, began to be colonized. John Mason's map of Newfoundland appeared in several publications in the 1620s, including William Vaughan's *Cambrensium Caroleia* (1625), while the Maritimes appeared on Sir William Alexander's *The Map and Description of New England* (1630), as propaganda for his proposed creation of Nova Scotia. Further south, Lord Baltimore's map appeared in *A relation of Maryland* (1635).

But besides the maps, partly propagandist and partly genuinely cartographical, the map printers were beginning to take a leading part in providing the context in which the new settlements were gradually establishing themselves. Thus John Speed's *Atlas* (1627), produced in England and depicting English settlements in America, provided the standard map of eastern North America for some years. But it was the Dutch professionals who best tried to keep up with the new developments, even if they were not above copying from each other, and were quite willing to present out-of-date material in their finely engraved and often lavishly decorated maps. To select an example or two, we might take Jan Jansson's *Nova Belgica et Anglia Nova* (Amsterdam, 1636), which is able to add effective representations of the Hudson River and much of New England, though much was copied from a comparable map by Joannes de Laet (Amsterdam, 1630). Willem Jansson Blaeu printed maps of parts of North America from 1617 onward, and in his *Atlas* of

1640 he produced a superior map of the English and Dutch areas. Nicholas Janz Visscher continued the development of the Anglo-Dutch area, adding in about 1655 a view of "Nieuw Amsterdam" and thus including for the first time a town plan on a topographical map. This was an idea that was widely copied later, as on Karel Allard's 1700 map of New England. It was maps like these that dominated much of the seventeenth-century cartography of eastern North America, assimilating, though slowly and often inaccurately, the expanding European presence in this area. In a different class were the Mercator projection sea charts of Robert Dudley. Those of North America in the third volume of *Dell'arcano del mare* (1661) represented careful study of all the manuscript and printed maps and charts Dudley could find, and their fine engraving and lettering set them apart from their contemporaries, even if the entirely original element in them is not large. John Farrer's manuscript and printed maps of 1650 and 1651 give much more detail for the Chesapeake area than Smith could attempt, but were quite unscrupulous about the context in which the genuine information was placed, the Pacific being shown at no great distance beyond the Appalachians, about which information was at that time very sketchy. Still, we have the general impression that after the middle of the seventeenth century there were large numbers of printed maps of North America in circulation. These must have been useful for navigators, greatly widening the perceptions of Europeans about what lay across the Atlantic Ocean.

After 1660, a series of English cartographers began to produce maps of the expanding British colonies, relying to some extent on original information. These cartographers form the so-called *Thames School,* a body of trained craftsmen who from early in the century were willing to produce to order, from manuscript or printed sources, charts or maps that were required for specific voyages. Their productions are easily identifiable by their style, which changed little during the seventeenth century and was workmanlike if not beautiful, and accurate within fairly circumscribed limits (they were producing plane, not Mercator, charts as late as 1700). Two men were particularly important in the production of engraved maps: John Thornton and William Morden. Thornton's *A New Mapp of the North Part of North America* and Robert Morden and William Berry's *A New Map of the English Plantations in America,* both published in 1673, set a standard that was gradually improved throughout the remainder of the century. There was no effective sea atlas for North America until John Thornton issued *The English Pilot. The Fourth Book* (1689); it had the best charts available at the time (in plane chart form), with sailing directions, and became a classic.

Under the auspices of William Blathwayt, a civil servant, the English Plantations Office had compiled before 1689 a remarkable collection of printed and manuscript maps of North America (and some other parts of the British empire), and this showed how effectively the Thames School sea charts could supplement what was already in print. *The Blathwayt Atlas,* as edited by Jeannette Black, is indeed a compendium of the late seventeenth-century cartography of North America as known to the English, just as Louis Hennepin's *Carte d'un Tres Grand Pais Nouvellement Découverte dans l'Amérique Septentrionale* (Amsterdam, 1697) summarizes the French cartographers' achievement of linking the Great Lakes with the Mississippi and the Gulf of Mexico.

If the English were slow to expand into the interior, they did continue to refine and illustrate in their maps the details of the settlement of individual colonies. Augustin Herrman (or Heerman) worked between 1659 and 1673 to refine a map of Maryland and much of Virginia, in the fine map eventually published in the latter year (Plate 2.8). Born in Prague, he had long served the Dutch West India Company, and in 1659 transferred his allegiance to Charles Calvert in Maryland, in exchange for denization and a grant of property. His map was published in London in 1673 in four sheets. It is one of the most carefully planned and constructed works of the period, though copies of it are exceedingly rare.

One might almost imagine that a final stage in the development of imperialist mapping of North America came with the engraving in Boston in 1677 of a detailed map of New England (Plate 2.9). Although this map owed something to English printed maps, it also contained the kind of detail that residents of the New England colonies would themselves wish to have at their disposal. But it was a flash in the pan, for well after 1700 the English and other European colonies in North America continued to depend most on maps and charts compiled and produced in engraved form in Europe. However, these maps remained relatively general in their coverage, and as we shall see in the next chapter, detailed large-scale maps of the newly settled areas hardly began to appear before the second half of the eighteenth century.

SOURCES FOR MAPS OF EUROPEAN EXPLORATION

GENERAL STUDIES OF NORTH AMERICAN CARTOGRAPHY

Cumming, William P.; R. A. Skeleton; and D. B. Quinn. *The Discovery of America.* New York: 1972. This volume consists essentially of texts concerning the discovery, but also contains good reproductions of many of the maps mentioned in this chapter.

Cumming, William P.; Susan Hillier; D. B. Quinn; and Glyndwr Williams. *The Exploration of North America.* New York: 1974. This is a continuation of the previously cited work by Cumming et al., and also contains many useful reproductions of maps.

Fite, Emerson David, and Archibald Freeman. *A Book of Old Maps.* (First published Cambridge:

1926; reprint available). Facsimiles of seventy-four maps, including many of the ones cited above, with a sound accompanying commentary.

Johnson, Adrian Miles. *America Explored*. New York: 1974). Subtitled *A Cartographical History of the Exploration of North America,* this is a fairly elementary account of the exploration and settlement of North America, illustrated with some unusual maps as well as the conventional ones.

Klemp, Egon. *America in Maps*. New York and London: 1976. Handsome facsimiles of important maps of the Americas, including some rarely seen ones from German libraries and archives.

Nordenskiöld, A. E. *Facsimile-Atlas*. Stockholm, 1889 (available in reprint); and Nordenskiöld. *Periplus*. Stockholm: 1887 (also available in reprint). These two books by the eminent Swedish explorer contain reproductions of many of the maps most important in understanding explorations in the New World, before and after 1492.

Quinn, D. B.; Alison Quinn; and Susan Hillier. *New American World: A Documentary History of North America to 1612*. 5 vols. New York: 1979. These volumes contain a series of black-and-white maps covering the period to 1612. There is also a brief accompanying commentary.

Schwartz, Seymour, and Ralph Ehrenberg. *The Mapping of America*. New York: 1980. A sound and comprehensive history of the cartography of North America, which of course extends far beyond the period of discoveries; useful color plates.

Shirley, Rodney W. *The Mapping of the World: Early Printed World Maps 1472–1700*. London: 1983. A comprehensive listing of world maps during the first two centuries of European expansion. It has full bibliographies and is invaluable for North America as well as for other regions.

DETAILED MONOGRAPHS

Black, Jeannette, ed. *The Blathwayt Atlas*. Providence: 1970 and 1975. This remarkable atlas offers a sort of conspectus of what an English official could assemble in the way of maps of North America at the end of the seventeenth century. It contains both printed and manuscript maps, and has a sprightly commentary.

Cumming, William P. *The Southeast in Early Maps*. Princeton, 1958 (2nd edition, revised and improved, 1962); and Cumming. *North Carolina in Maps*. Raleigh: 1966. These two volumes cover the southern half of the eastern seaboard with a sure commentary and fairly large-scale reproductions of the more significant maps.

Hulton, Paul, ed. *America 1585: the Complete Drawings of John White*. Raleigh, N.C.: 1984. A readily available collection of John White's drawings, with 106 black-and-white illustrations and 77 color plates. Hulton has also edited:

• (with David Quinn) *The American Drawings of John White, 1577–1590*. 2 vols. London: 1964.

• Thomas Harriot. *A Briefe and True Report of the New Found Land of Virginia*. New York: 1972.

• *The Work of Jacques Le Moyne de Morgues, A Huguenot Artist in France, Florida and England*. 2 vols. London: 1977.

Lorant, Stefan, ed. *The New World: The First Pictures of America*. Philadelphia: 1965. A set of reproductions of Theodore de Bry's engravings and the watercolors of John White. It is widely available, but should be used with great caution as the plates and text were much criticized at the time of its publication.

Mollat du Jourdin; Michel; and Monique de la Roncière. *Sea Charts of the Early Explorers*. London: 1984. A splendidly produced volume, covering the period up to 1700. Many of the examples are taken from the collections of the Bibliothèque Nationale in Paris. There is of course no specifically North American emphasis, but there are in fact a good many maps showing that part of the world, and they are very well reproduced.

Rotz, Jean. *The Boke of Idrography,* edited by Helen Wallis. London: 1981. This splendid atlas by Jean Rotz, offered to Henry VIII of England in 1542, offers us a summary of what a leading cartographer of the Dieppe School knew in the middle of the sixteenth century. The reproductions are superb, and the commentaries very instructive.

Skeleton, R. A. *Explorers' Maps*. London: 1958. This book, subtitled *Chapters in the Cartographic Record of Geographical Discovery,* contains some very interesting sections on North America.

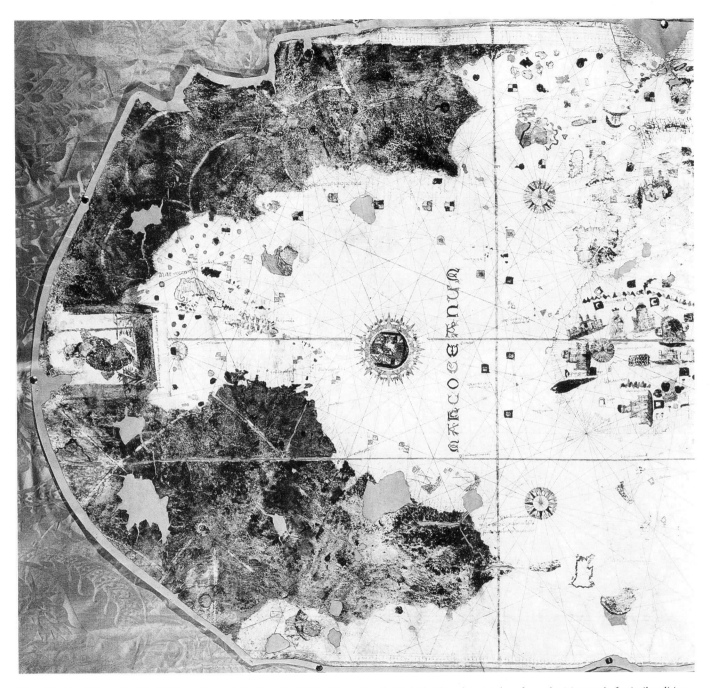

Plate 2.1. Detail from the World Map of Juan de La Cosa (c. 1500). Museo Naval, Madrid. This plate is taken from the Museum's facsimile edition.

Plate 2.1a. Diagram of Plate 2.1.

An Early European Cartographic Image of the New World

This marvelous map, which was lost between 1500 and 1832, is the first to show the discoveries of Columbus. It was drawn by Juan de la Cosa, who almost certainly accompanied Columbus on one or more of his voyages, and is a very mine of information, much of it hard to interpret.

To the west we see a rectangular frame, within which is the figure of Saint Christopher, who no doubt stands for Columbus, the *Christoferens* bearing Christianity to the pagans. This frame has the advantage of covering a section of the map open to doubt, for it was not yet known what lay to the west of Cuba: a short stretch of water before the coast of Asia was reached, perhaps.

The islands of the West Indies are in general well delineated, for it was relatively easy to plot their general outline and position, even from a small boat in rough seas. Cuba is

shaped rather like a scorpion, a distortion that would last for a decade or two. Notice that the flag of Castille flies over the Caribbean area, but that to the north there are four English flags in the area marked "mar descubierta por los Ingleses" ("sea discovered by the English"), no doubt in reference to the voyages carried out in 1497 and 1498 by the Venetian John Cabot, on behalf of Henry VII of England, but here much extended.

The mainland of what is now South America is shown in a very distorted way, perhaps as a faint echo of some voyages made there about 1500. The whole mainland is covered by portions of circles, with what look like lakes towards their centers; it is not known what these represent. The map is technically not very sophisticated; it has for instance no indication of latitude, and of course none of longitude. The right-hand side of the map offers us an image of Europe, Asia, and Africa in the purest portolan chart style, with exaggerated embayment of the coasts, a very elaborate wind rose, and the use of flags to indicate territory. The left-hand side has these features in slightly less pronounced form, but the map as a whole remains firmly within the medieval portolan chart tradition. It was, indeed, largely with medieval tools that the secrets of the New World would be unlocked.

Sources: Emerson David Fite and Archibald Freeman. A Book of Old Maps *(first published Cambridge: 1926; reprint available); and Michel Mollat du Jourdin and Monique de La Roncière,* Sea Charts of the Early Explorers *(London: 1984).*

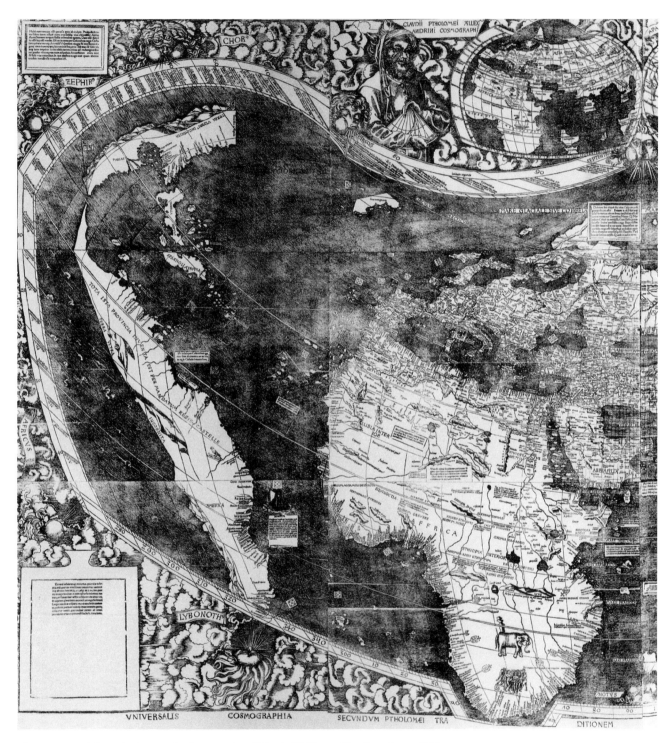

Plate 2.2. Detail from the *Universalis Cosmographia* of Martin Waldseemüller (Strassburg, 1507).
This plate was taken from Fite and Freeman, *A Book of Old Maps*.

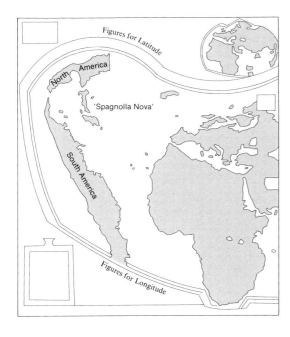

Plate 2.2a. Diagram of Plate 2.2.

A New Continent Emerges in Printed Maps

Until 1506, the year of his death, Columbus believed that the new-found lands were part of Asia, as did most of his contemporaries. However, the continuing explorations of Spaniards, Portuguese, and Englishmen began to suggest to some scholars that these very extensive new lands had to be part of a new continent, and this map by Martin Waldseemüller was among the first to express this idea in cartographic form.

Waldseemüller was a member of a small group of cosmographers working in Lorraine, often under the patronage of the Duke of Lorraine. He drew several maps, and this one was among the most popular. One thousand copies of it were sold, but none of them was known to survive until this unique copy was found in 1901 in the library of Prince Waldburg-Wolfegg in Württemberg. Waldseemüller relied heavily on the publications of Amerigo Vespucci and used his name for the more south-

erly of the two continents that he posited west of the Caribbean.

The delineation of the southern continent is very sketchy, and of course the whole of its west coast is an inspired guess, for it was not until 1513 that Balboa crossed the isthmus of Panama. Moreover, the shape of the east coast of North America and of the Gulf has to be conjectural, for the Gulf was not mapped before 1519, and much obscurity surrounds the various supposed voyages up the east coast before 1510.

Waldseemüller's map is in some respects a great advance over that of Juan de la Cosa; in particular, he took care to tie his new world in with the old world by a systematic grid of latitude and longitude, quite lacking on the la Cosa map. Even though some of the positions were wrong, this attempt to introduce mathematical rigor laid the foundation for future improvements. We might say that with its theoretical background and systematic projection this map is characteristic of the work of early modern theorists, as opposed to the empirical seaman's style of the la Cosa map.

Sources: Emerson David Fite and Archibald Freeman, A Book of Old Maps *(Cambridge: 1926; reprint available); and Seymour Schwartz and Ralph Ehrenberg,* The Mapping of America *(New York: 1980).*

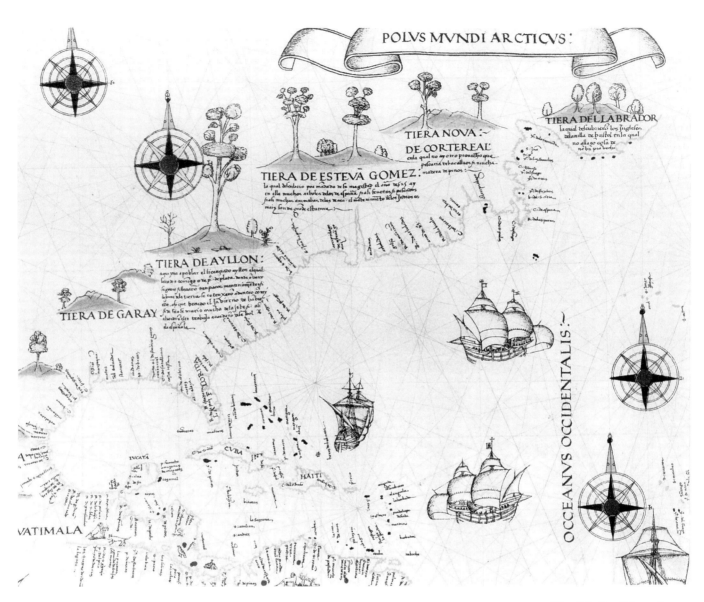

POLVS MVNDI ARCTICVS :

TIERA DELLABRADOR
la qual tefcubrieron los yngleſes.
adaniſla de briſtol enla qual
no alla eo coſa de
neba pue bedia

TIERA NOVA :·~
DE CORTEREAL
enla qual no ay o tro pronecho que
peſcaria trbacallaoſ ſi mucha
madera de pinos :·~

TIERA DE ESTEVĀ GOMEZ :
la qual dlcubrio por mado de ſu mageſtho el año de25 ay
en ella muchos arboles delos de eſpiña paſi fructas peſcaeſ
ſeſi muchas animalias delas vaca el mātenimiēto delos Judios es
maiz ſon de grde eſtatura :

TIERA DE AYLLON :
aqui yua apoblar el licenciado ayllon elqual
ſalio a comigo o de p deplata donte aben
ſigente ſilen oto tan poca mantnimꝰtoyſi
lebno ala terra ſe retos xero aduntro como y
oto a que benuo el ſn dierno de babie
ſto fuo ſe murio mucha delaſ te oſi ali
eba evo eſte trabajo acuadero de bol X
de eſpanola

TIERA DE GARAY

OCCEANVS OCCIDENTALIS :

VATIMALA

IVCATĀ

CVBA

HAITI

Plate 2.3. Detail from the
World Map of Diogo Ribeiro
(1529). Vatican Library.
Reproduced from the
published facsimile edition.

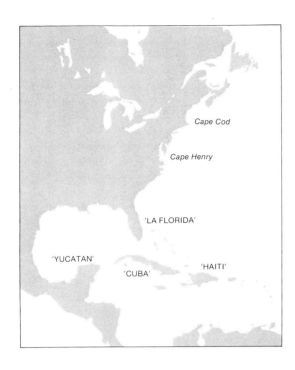

Plate 2.3a. Modern outline of North America.

The Spanish Cartographic Vision of the World in 1529

The Spanish Crown organized its overseas ventures with an all-encompassing thoroughness that included the formation of a navigation school at Seville, with pilots general whose job it was to keep the master map, or *padrón real,* up-to-date. Most of the maps deriving from the padrón real have perished, but about half a dozen survive, and on them we can see the world taking shape for the Spaniards with marvelous fidelity. The map from which a detail is shown here is the most remarkable of the surviving copies; extraordinary not only for the range of its cartographical innovations, but also for its great beauty. It is now in the Vatican Library in Rome, and has been there since the Emperor Charles V presented it to Pope Clement VIII in 1529.

Its author was the royal cosmographer Diogo Ribeiro, and in presenting it to the Pope the Emperor no doubt wished to substantiate Spanish claims to wide areas of the New World. By the Treaty of Tordesillas, 1494, Spain received papal recognition of her claim to all lands west of a meridian (north-south line) drawn 370 leagues (roughly 1,175 miles) west of the Cape Verde Islands. Of course, the difficulties in establishing longitude made it almost impossible to be sure in 1529 just where this line ran in relation to many areas of the New World; in fact, it falls about one hundred miles east of Rio de Janeiro.

The east coast of North America is probably the weakest of the regions shown on the Ribeiro map, because by the late 1520s Spanish attention was increasingly focused on Central and South America. Because accurate latitude observations were not available, the discoveries of Gomez were brought too far down the coast, one result being that Cape Cod was moved so far south that it might appear to be correlated with Cape Henry or even Cape Fear. In fact, the whole east coast is excessively elongated towards the east; half a century would elapse before this area would be mapped with any accuracy.

Source: William P. Cumming, R. A. Skelton, and D. B. Quinn, The Discovery of North America *(London: 1971).*

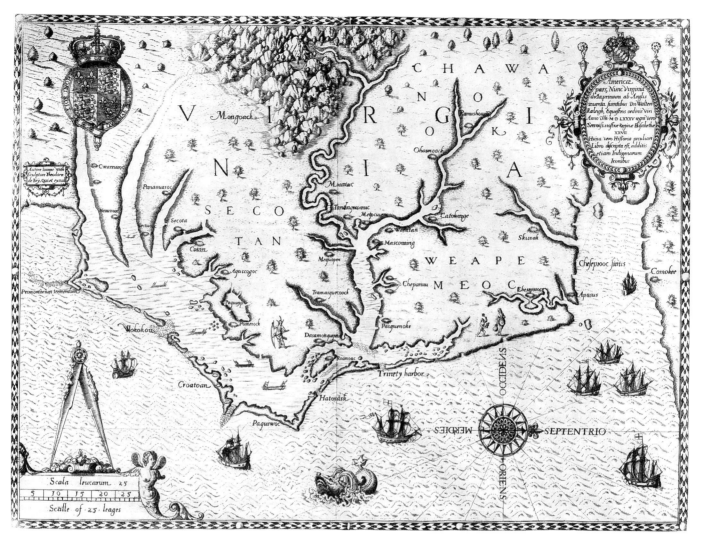

Plate 2.4. Virginia, from Theodore de Bry's *America* (13 parts, Frankfurt, 1590–1634).

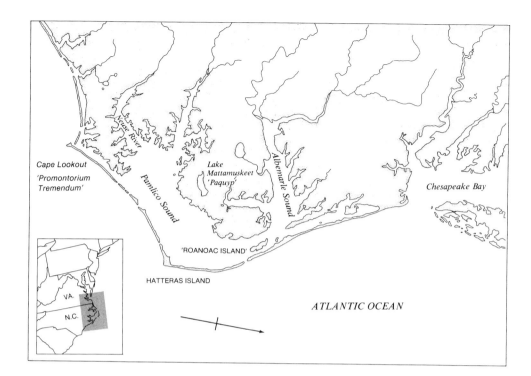

Plate 2.4a. Modern map covering area shown in Plate 2.4.

Cape Lookout
'Promontorium
Tremendum'

Neuse River

Lake
Mattamuskeet
'Paquyp'

Albemarle Sound

Chesapeake Bay

Pamlico Sound

'ROANOAC ISLAND'

HATTERAS ISLAND

ATLANTIC OCEAN

VA.

N.C.

A Large-Scale English Map of the 1580s

This map from Theodore de Bry's compendious *America* (Frankfurt 1590) essentially reproduces the manuscript map of the area surrounding the Carolina Sounds produced by John White and Thomas Harriot in 1585–1586, during the Roanoke Island colonizing venture organized by Sir Walter Raleigh.

White and Harriot were a remarkable team, and their work is exceptionally accurate, as may be seen from our sketch of a modern map of the same area, now mostly in North Carolina (Plate 2.4a). It is only in the extreme north, around the Chesapeake Bay, and in the south, around the Neuse River, that their outline breaks down; here no doubt they were at the end of their accurate survey, and simply giving an impression of what they saw.

One of the most interesting features of the map is the abundance of Indian sites. Very few of them have as yet been identified by ar-

chaeology, and some indeed may by now have been destroyed by the sea, but White gives us a good idea both of their location and of their names. The only unsatisfactory feature of the engraved map is the mountain range to the east (top center), and this seems to have been added as a fantasy of the engraver, or perhaps to indicate the possibility of gold.

The map is also interesting for its symbolic content. Although many Indian sites are marked, with small stockades, these are very much smaller than the circling English ships. One has the impression, looking at the map, that a conquering people is coming to take over an almost empty country, and this fantasy later in fact became the European justification for the invasion of America.

Source: William P. Cumming, The Southeast in Early Maps *(Chapel Hill, 1962).*

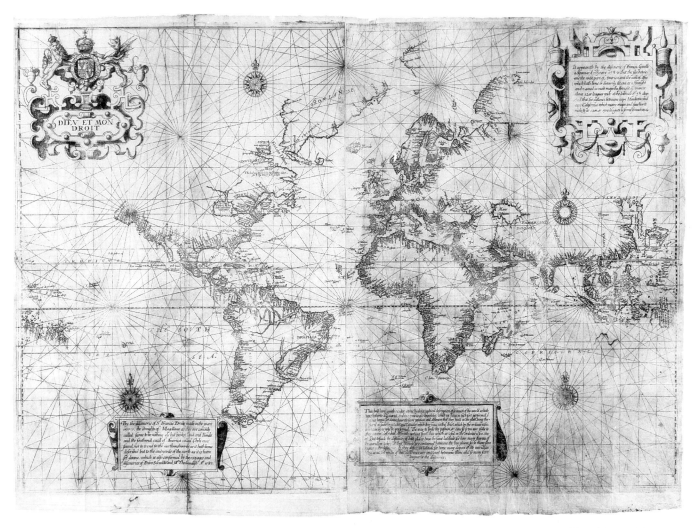

Plate 2.5. World Map by
Edward Wright (London,
1599).

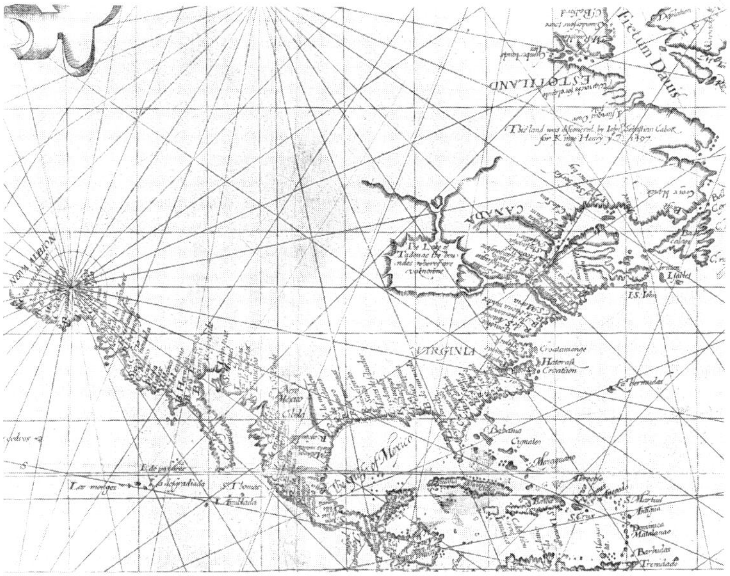

Plate 2.5a. Detail from Wright's World Map.

An English Synthesis of Cartographic Knowledge in 1599

This map offers the best synthesis of knowledge of North America that could be obtained in England at the end of the sixteenth century. It is now known to be the work of Edward Wright, and is constructed on the Mercator projection, Wright having been the first person (in his *Certain Errors in Navigation*, 1599) to give a mathematical model for the technique of drawing a Mercator chart. He had assisted in the compilation of Emery Molyneux's globe of 1592, and was well informed on American geography. The lower cartouche reads: "Thou hast here (gentle reader) a true hydrographical description of so much of the world as has beene hetherto discovered and is comme to our knowledge," and the map does give a remarkably accurate account of what was known. Even more remarkable for the period, it resists the temptation to insert fantastic coastlines where the cartography was still in doubt.

Note that Wright has some information concerning the Great Lakes, which he describes as "The Lake of Tadouac, the boundes whereof are unknown" (Plate 2.5a). Note, too, that he does his best to substantiate English claims, in the North to the land discovered by the Cabots, and in the far West to Drake's "Nova Albion." This map is very rare, and came originally from volume 2 of the second edition of Richard Hakluyt's *The Principal Navigations, Voyages, Traffiques, and Discoveries of the English Nation, made by Sea or overland, to the remote and farthest distant quarters of the earth* (London: 1599). Just as Hakluyt's great work summarizes in written form what the English have achieved to date, and offers inspiration for the future, so this map gives us a compendium of cartographic knowledge to that date, and would be of use to intending colonists; it was, for instance, undoubtedly known to the founders of Jamestown in 1607.

Sources: Emerson David Fite and Archibald Freeman, A Book of Old Maps *(Cambridge: 1926; reprint available); and Rodney W. Shirley,* The Mapping of the World: Early Printed World Maps 1472–1700 *(London: 1983).*

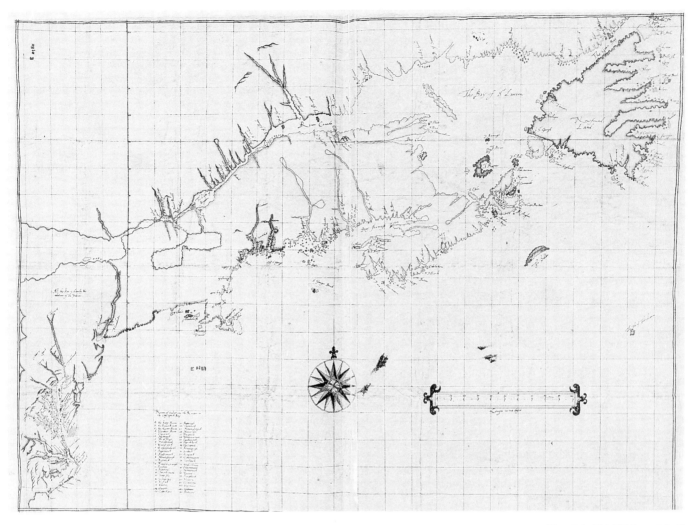

Plate 2.6. The "Velasco Map" (1610). Archivo General de Simancas, Spain.

David B. Quinn

An English Summary of Discoveries in the Northeast, 1611

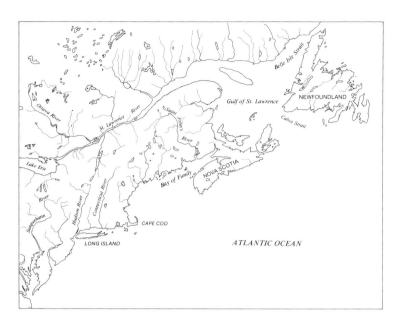

Plate 2.6a. Modern map covering area shown in Plate 2.6.

This map, now in the Spanish Archivo General at Simancas, is the result of much international espionage. It was sent back to Spain in February 1611 by the Spanish ambassador in London, Alonso de Velasco, who had been there since 1609 and would leave in 1613. He surely must have obtained it by subterfuge, for it contains much confidential information, not only from English sources but also from French ones, the latter no doubt obtained surreptitiously by the English.

The map is a compendium of discoveries in Northern American waters between 1584 and 1610. The most southerly section is based on John White's map of the territory explored by the English between 1584 and 1586, while the Virginia area comes from a manuscript map by Captain John Smith that would not be published until 1612. The coast between Cape May and modern Connecticut comes from the manuscript of Henry Hudson's mapping there in 1609, and the New England coast seems in part to be a conflation of lost English maps of 1602, 1605, and 1607–1608.

The other source for the New England coast seems to be the work of Samuel de Champlain between 1604 and 1607, unpublished but represented by a manuscript map now in the Library of Congress.

Other information deriving from the French must account for the outline of Newfoundland, the Gulf of the Saint Lawrence, the course of the river as far as Montreal, and the conjectural extensions of the Richelieu River and of the Saint Lawrence River into unexplored lakes; this information would not be published before 1613.

Spanish ambassadors in England were of course always trying to obtain secret information about new exploration, and Velasco's predecessor had laid hands on an unpublished map of Virginia. Moreover, all the European powers did their best to obtain information from the Spanish cartographic center at Seville. Still, this map obtained by Velasco is exceptional in the range, novelty, and accuracy of the information that it contains.

Sources: William P. Cumming, R. A. Skelton, and D. B. Quinn, The Discovery of North America *(London: 1971) (fine color reproduction); and Emerson David Fite and Archibald Freeman,* A Book of Old Maps *(Cambridge: 1926; reprint available).*

AUGUSTANA UNIVERSITY COLLEGE LIBRARY

Plate 2.7. *Virginia* by John Smith (London, 1612).

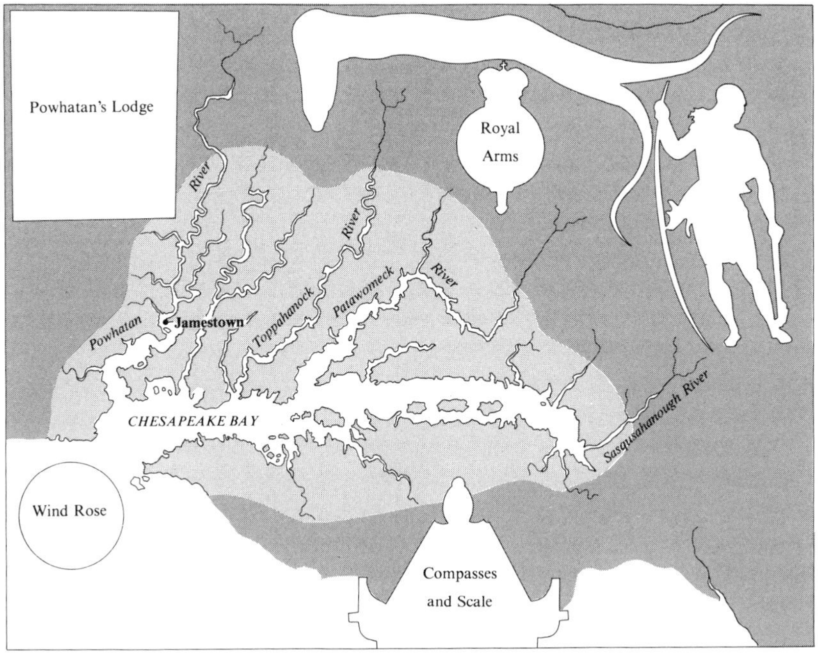

Plate 2.7a. Diagram of Plate 2.7.

Powhatan's Lodge

River

Royal
Arms

Powhatan

Toppahanock

River

Patawomeck

River

River

Jamestown

CHESAPEAKE BAY

Sasquahanough River

Wind Rose

Compasses
and Scale

An Early English Image of the Newly Settled Land

Captain John Smith compiled this map during his explorations of the Chesapeake Bay between 1607 and 1609; it was published in England in 1612, perhaps in spite of the opposition of the Virginia Company, which had become hostile towards Smith and was in any case secretive about their colony. This map represents an extraordinary amount of activity for a single person. Smith was unusually conscientious in indicating the limits of his personal knowledge by Maltese crosses; the areas beyond were drawn from hearsay information from Indians, or reports of excursions by colonists. He also plotted a great many Indian villages, the accuracy of which modern archaeology has tended to confirm, just as later cartography shows how well he drew the rivers with the means at his disposal (Plate 2.7a).

The only major weaknesses in the delineation are that the Chesapeake Bay is too narrow and its islands shown in summary fashion;

Smith also probably knew more about the coast of the Virginia eastern shore than he put on the map. He copied the decoration in the upper-left corner from the engraved pictures of Indian life published in 1590 by Theodore de Bry (and drawn by John White), thus beginning a long tradition of plagiarism in colonial iconography.

This map went through many editions and was no doubt widely known in England. It enabled intending colonists to plan future settlements, especially from 1618 onward, and was helpful for merchants organizing trading expeditions to specific areas; without a map of this kind their task would have been much more difficult.

Sources: Seymour Schwartz and Ralph Ehrenberg, The Mapping of America *(New York: 1980); and Edward Papenfuse and Joseph Coale III,* Atlas of Historical Maps of Maryland, 1608–1908 *(Baltimore: 1982). The* Atlas *has a particularly interesting comparison between the Smith map and a recent satellite image.*

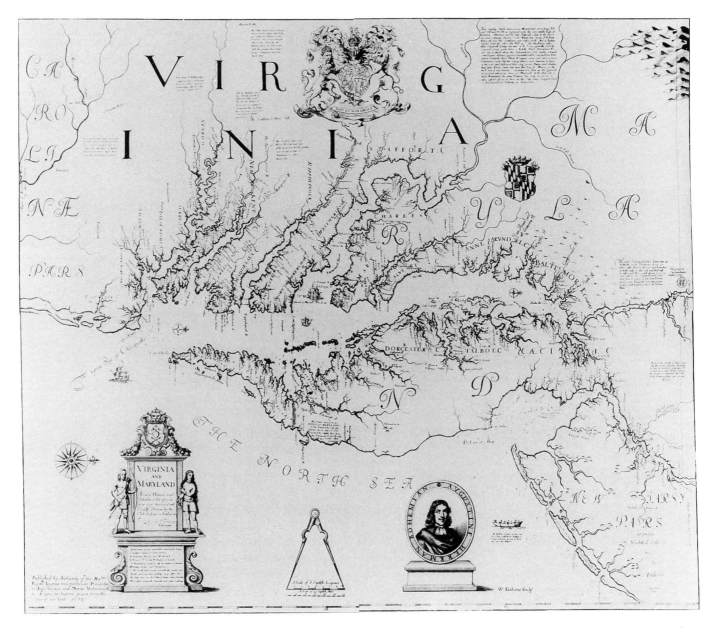

Plate 2.8. *Virginia and Maryland* by Augustine Herrman (London, 1673).

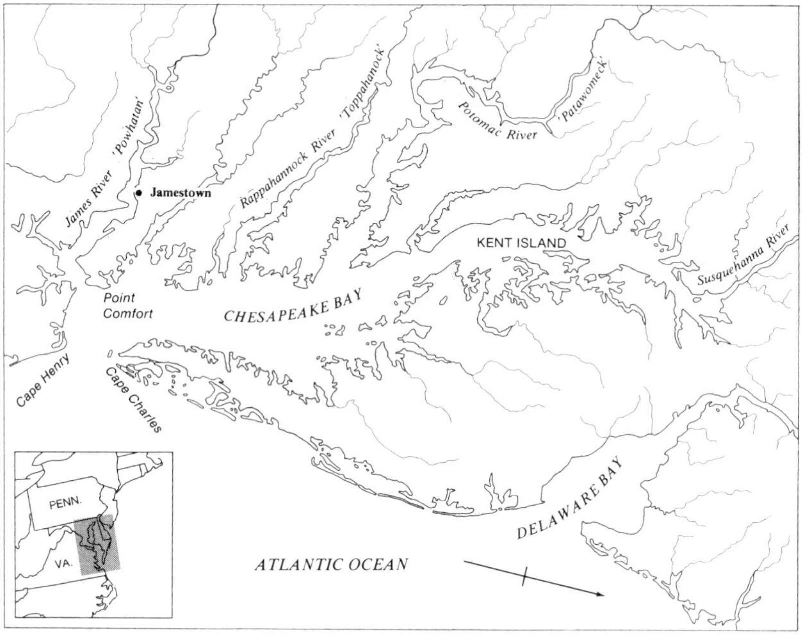

Plate 2.8a. Modern map covering the area of Plate 2.8.

Map from the Stage of European Consolidation

Augustine Herrman had come into the service of the governor of Maryland after long years with the Dutch West India Company. He was therefore particularly concerned to define the Dutch settlement on the Delaware and the borders of the Maryland portion of the Eastern shore, together with the contested Kent Island in the bay, but he was as well a very gifted topographer, whose appreciation of the physical detail and of the location of settlements makes his map one of the most remarkable of the period.

The coastal detail would not be surpassed for more than half a century, and it would seem that all the places occupied by the time of the map's completion were named and located. The map's only weakness is that Herrman did not penetrate into the interior, and so could not fully define the territory claimed by the Calverts. In a wider sense, too, Herrman shared the general ignorance of the English sea-board settlers; he thought that the Spaniards were just across the mountains, but knew nothing of the French explorations, which that very year would lead to the first sure European knowledge of the Mississippi River.

It is instructive to compare Herrman's map with that of Smith, drawn half a century earlier (Plate 2.7). The "Powhatan" has become the James River, but most of the other Indian river names have been only slightly adapted: "Toppahanock"—Rappahannock River, "Patawomeck"—Potomac River, and so forth. The area of settlement has much increased, but it remains confined to the banks of the rivers; there were very few internal roads at this time, and the economic life of the surrounding communities remained essentially waterborne.

Sources: William P. Cumming, S. E. Hillier, D. B. Quinn, and G. Williams, The Exploration of North America 1630–1776 *(New York: 1974); and Emerson David Fite and Archibald Freeman,* A Book of Old Maps *(Cambridge: 1926; reprint available).*

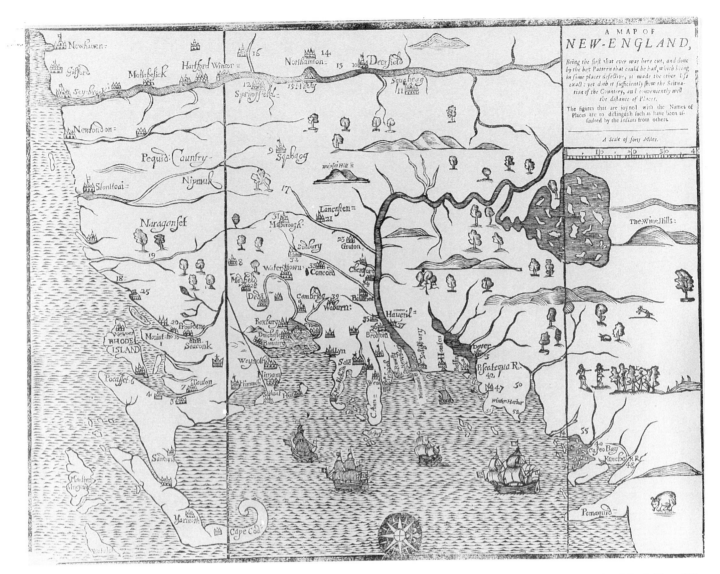

A MAP OF
NEW-ENGLAND,

*Being the first that ever was here cut, and done
by the best Pattern that could be had, which being,
in some places defective, it made the other less
exact: yet doth it sufficiently shew the Scituation of the Countrey, and conveniently well
the distance of Places.*

*The figures that are joyned with the Names of
Places are to distinguish such as have been as-
faulted by the Indians from others.*

A Scale of forty Miles.

Plate 2.9. *A Map of New
England,* by John Foster
(Boston, 1677).

Plate 2.9a. Modern map covering the area of Plate 2.9.

The First Printed Map from English North America

This is the first printed map made in the English colonies, and was probably produced by John Foster, publisher of William Hubbard's *A Narrative of the Troubles with the Indians in New England* (Boston: 1677). The primary aim of the map is to show the widespread nature of Indian risings described in Hubbard's book; the numbers on the map represent places where incidents with the Indians had occurred, some being attached to names and some not.

The two vertical black lines indicate the boundaries of the Commonwealth of Massachusetts, with no such indication for New Plymouth, Rhode Island, or parts of Connecticut and Maine. Although the map is technically rather crude, using the by now outdated woodcut process, it draws heavily on information provided by recent surveys by the government of Massachusetts, and so provides much accu-rate detail, particularly in locating individual settlements and in giving the precise locations of areas affected by hostile Indian activity. For a pioneer effort it is a considerable achievement, and added greatly to the value of and interest in Hubbard's work.

Note the inscription below the title, accurately describing the map's qualities and defects: "A map of New England, being the first that ever was here cut, and done by the best pattern that could be had, which being in some places defective, it made the other less exact: yet doth it sufficiently shew the scituation of the countrey, and conveniently well the distance of the places."

Sources: Emerson David Fite and Archibald Freeman, A Book of Old Maps *(Cambridge: 1926; reprint available); and Seymour Schwartz and Ralph Ehrenberg,* The Mapping of America *(New York: 1980).*

CHAPTER THREE

Eighteenth-Century Large-Scale Maps

Louis DeVorsey

It is a daunting task to give an overview of large-scale maps (that is, maps showing a relatively small area in some detail) of eighteenth-century North America. At the beginning of the eighteenth century the strip of settlement along the eastern seaboard was still only slenderly established, and there were great areas of wilderness separating the isolated European outposts. As the century continued, the areas of wilderness became smaller and the settlements steadily grew, until by the 1770s most of the best land between the Appalachians and the sea had been taken up, and a great mass of people was poised for the next stage, which was the European invasion of the North American heartland. The maps kept pace with this extraordinary expansion, both in the sense of offering at each stage general images of the extent of settlement, and in the sense of providing detailed coverage—now usually drawn by native cartographers—of the newly settled lands.

The achievements of the first third of the eighteenth century were summarized in Henry Popple's *A Map of the British Empire in America,* published in London in 1733 (Plate 3.1). This huge twenty-sheet map, which when assembled measured about eight feet square,

brought together all the best sources, both French and English, to give an extraordinary image of North America after the first century of European settlement there. The cartographic mastery of the area east of the Mississippi River is impressive, for the course of the great river had now been well delineated, and so had its relationship to the Great Lakes. However, when we look closely at the place-names, we are reminded of how slow the eastern settlements were to take off, demographically speaking, for there still were no towns of any size away from the sea and tidewater country.

About twenty years later, in 1755, John Mitchell published a sort of update of Popple, entitled *A Map of the British and French Dominions in North America.* This very influential map was still consulted four decades later, and was the cartographic document upon which Great Britain and the United States based their claims in the negotiations leading up to the treaty that terminated the Revolutionary War. It showed us the area on the eve of the Seven Years' War, a war that would result in a radical change in the balance of power, with the French eliminated as a political force. It would be interesting to make a close comparison between Popple and Mitchell for certain chosen areas, to determine whether the latter showed any greatly increased density of settlement. In fact, both Popple and Mitchell may be used either as a whole, to show the existing extent of cartographic knowledge, or in detail, to point up the stage reached by some nodal area like the Chicago portage (Plate 3.1).

After the peace of 1763, the colonies on the eastern seaboard began a period of rapid growth, and their cartography kept pace with this development. Following their victory, the British appointed Samuel Holland Surveyor General for the area north of the Potomac River, and William Gerard De Brahm his counterpart for the area south of that. Holland produced maps for New York and New Jersey, and De Brahm was active in Florida (see Plate 3.2). Other areas too were covered at this time of great cartographic development; there was a map by William Scull for Pennsylvania, by John Fry and Peter Jefferson for Virginia and Maryland, and by Henry Mouzon for the Carolinas (Plate 3.3). These maps are on a relatively large scale, and of course they now show a greatly intensified rhythm of settlement, which in many cases laps up against the line of the Appalachian Mountains.

Many of these maps appeared in the great atlases that were characteristic of the period, such as Thomas Jefferys' *A general topography of North America* (London, 1768) and Jefferys' *The American Atlas* (London, 1776); and William Faden's *The North American Atlas* (London, 1777). Some of this material is readily available in facsimile, for Jefferys' *The American Atlas* was reprinted in 1974, and *North America at the Time of the Revolution* brings together many of the state maps (see the bibliography for this chapter for details of this publication). Moreover, in 1972 the Naval History Division published *The American Revolution 1775–1783: An Atlas of 18th Century Maps and Charts,* which is another useful compendium of large-scale maps, well reproduced.

These detailed topographical maps are very useful for investigating local problems and identifying particular places; for instance, the Mouzon detail (Plate 3.3) can be used to locate the area covered by the estate plan reproduced as Plate 3.5. They are also useful for imaginatively reconstructing the landholding pattern of large areas, giving us, for instance, an image of the great estates along the banks of the Ashley and Cooper Rivers (Plate 3.3), and can be used more prosaically for establishing the lines of roads and the boundaries of newly established counties.

The Revolutionary War gave rise to a great spate of mapmaking, not only by commanders in the field, but also by English publishers, who wanted to give their readers an idea of the nature of operations in that distant theater. These developments have been described in *Mapping the Revolutionary War* (Chicago, 1978), by J. B. Harley, Barbara Petchenik, and Lawrence Towner. The "magazine map" was another type produced in great numbers at this time; the middle of the eighteenth century was the heyday of periodicals such as *The Gentleman's Magazine* and *The London Magazine,* and these reviews often carried detailed maps of events in North America as well as elsewhere in the world. The maps were an important part of the selling appeal of these publications, and could be engraved and colored in great numbers, which seems almost impossible to us, given the absence of mechanization. The periodicals may be found in many libraries, and the maps may be tracked down using Christopher Klein's *Checklist* (see the bibliography at the end of this chapter).

Town plans were among the earliest maps drawn in the colonies. Whereas field boundaries could remain more or less fluid for long periods without undue inconvenience, it was important from the start to define urban property lines. The mapping of New York City was covered in the classic work by I. N. Phelps Stokes, *The Iconography of Manhattan Is-*

land (6 vols., New York, 1915–1928), and many remarkable views of early New York City may be found here, but on the whole the other eastern cities lack works of this kind. The best way to find out what exists in the way of colonial-period town plans is to use the numerous works of John Reps, listed in the bibliography for this chapter. Particularly noteworthy is his *Tidewater Towns: City Planning in Colonial Virginia and Maryland* (Williamsburg, 1972), which uses large-scale maps as well as town plans in its masterly reconstructions. Such plans can be remarkable teaching instruments, for students can be shown how to peel back successive layers of some city with which they are familiar, until they reach—as all will reach—the period when only Indians were settled there.

Maps of the countryside are harder to find. They range in complexity from simple geometrical diagrams, designed to set out the corners of a holding, to full-blown estate plans, of the type shown on Plate 3.5. The incidence of these estate plans in North America is interesting. They are not to be found where the proprietors were of modest means, or where some general cartographic system had been established. We thus look in vain for them in New England, where the farms were mostly small, or in the lower Mississippi valley, where an excellent cadastral survey was early established. They are on the other hand abundant in South Carolina, where the plantocracy held broad acres, and there was no satisfactory general survey.

These estate plans are of great value in reconstructing a past countryside. From them we can discover not only what crops were grown, and in what fields with what names—"Mill Field," "Bottom Field," and so forth—but also where the people lived, and how they moved about the countryside. We can find out where the mills were for extracting sugar and carrying out other processes, and sometimes we can locate them by archeology or by simple survey. After 1785, much of the country came to be covered by the township and range system, and this too offers extraordinary possibilities for reconstructing the countryside; the material surviving from this latter cartographic system is so complicated and so abundant that we have left it to the next chapter.

SOURCES FOR EIGHTEENTH-CENTURY LARGE-SCALE MAPS

Brown, Lloyd Arnold. *Early Maps of the Ohio Valley: a Selection of Maps, Plans and Views made by Indians and Colonials from 1763 to 1783.* Pittsburgh: 1959. About fifty maps of the Ohio Valley, with a very perceptive commentary.

Cumming, William P. *The Southeast in Early Maps.* Chapel Hill: 1962. A slightly more numerous collection of maps, both printed and manuscript, with an illuminating commentary. It is a pity that in both Cumming's and Brown's collections of maps the commentary is so far separated from the maps themselves.

Cumming, William P.; Susan Hillier; D. B. Quinn; and Glyndwr Williams. *The Exploration of North America.* New York, 1974. This handsomely illustrated volume carries the story of exploration and indeed of settlement down to 1776.

DeVorsey, Louis. *The Indian Boundary in the Southern Colonies, 1763–1775.* Chapel Hill: 1966. This book points out that maps showing the countryside should include reference to boundaries both European and European-Indian.

————, ed. *De Brahm's Report of the General Survey in the Southern District of North America.* Columbia: 1971.

Fite, Emerson David, and Archibald Freeman. *A Book of Old Maps.* First published Cambridge: 1926; reprint available. The final thirty maps in this volume concern the theme of this chapter. The facsimiles are well reproduced, with a facing commentary.

Graffagnino, J. Kevin. *The Shaping of Vermont: From the Wilderness to the Centennial, 1749–1877.* Rutland/Bennington: 1983. The first half of this book, which deals with thirty-five maps in all, covers eighteenth-century large-scale maps. Some of the reproductions are in color, and the discussion of the maps is enlivened by other visual material.

Hudson, Charles M., ed. *Red, White and Black: Symposium in Indians in the Old South.* Athens: 1971. Contribution by Louis De Vorsey on "Early Maps as a Source in the Reconstruction of Southern Indian Landscapes."

Jefferys, Thomas. *The American Atlas.* Walter Ristow, ed. Amsterdam: 1974. This collection of late eighteenth-century large-scale maps has an interesting introduction and a useful bibliography.

Jolly, David C., ed. *Maps of America in periodicals before 1800.* Brookline, Mass.: 1989. This useful list contains references to about 450 maps.

Klein, Christopher, M., ed. *Maps in Eighteenth-Century British Magazines.* Chicago: 1989. Another listing of maps in periodicals, this time covering the whole world.

Martin, James C., and Robert Sidney Martin. *Maps of Texas and the Southwest, 1513–1900.* Albuquerque: 1984. This collection of about fifty facsimiles of maps of the southwest contains about half a dozen concerning our period, and a useful bibliography.

Naval History Division, *The American Revolution 1775–1783: An Atlas of 18th Century Maps.* Washington, D.C.: 1972. The title is slightly misleading, for this in fact is a fine collection of eighteenth-century large-scale maps showing the area where the revolutionary war would be fought; it comes with a 60-page gazetteer of place-names.

North America at the Time of the Revolution. 3 parts. Lympne, England: 1972, 1974, and 1975. This three-part collection of facsimiles, with introductions, is divided as follows:

Part I, edited by William P. Cumming and Helen Wallis, contains only Popple (1733).

Part II, edited by Louis DeVorsey, contains Mitchell (1755), Fry and Jefferson (1775), Sauthier (1776), Holland (1776), and Jefferys on New England (1774).

Part III, edited by William P. Cumming and Douglas W. Marshall, contains Bowen and Gibson (1775), Scull (1775), Mouzon (1775), Ross (1765), Jefferys on West Florida (1775),

Jefferys on Nova Scotia (1775), and Jefferys on the Saint Lawrence (1775).

The three parts of this publication together make up an excellent sample of eighteenth-century large-scale maps, well reproduced at a reasonable price.

Papenfuse, Edward C., and Joseph M. Coale III. *Atlas of Historical Maps of Maryland, 1608–1908.* Baltimore: 1982. These maps are exceptionally well reproduced and have a lively commentary. The first 50 of the 150 apply to the period before 1800.

Proctor, Samuel, ed. *Eighteenth-Century Florida and its Borderlands.* Gainesville: 1975. Contains De Vorsey. "DeBrahm's East Florida on the Eve of Revolution."

Reps, John. *Cities of the American West: A History of Frontier Urban Planning.* Princeton, N.J.: 1979. A great compendium of views, both manuscript and printed, many of them dating from the eighteenth century, together with a narrative.

————. *The Forgotten Frontier: Urban Planning in the American West before 1890.* Columbia: 1981. A shorter, self-contained version of *Cities of the American West.*

————. *Tidewater Towns: City Planning in Colonial Virginia and Maryland.* Williamsburg: 1972. Contains plans and much other material concerning the early cities of the east. Many of the plan sequences would make excellent teaching aids.

————. *Town Planning in Frontier America.* Princeton: 1969. This compact book is based largely on the author's earlier *The Making of Urban America* . (1965). Its format is rather small, but it contains a wealth of town plans from the eighteenth century and earlier.

Ristow, Walter. *American Maps and Mapmakers: Commercial Cartography in the Nineteenth Century.* Detroit: 1985. In spite of its subtitle, this book's early chapters contain much about the large-scale maps of the eighteenth century.

————, ed. *A la Carte: Selected Papers on Maps and Atlases.* Washington, D.C.: 1972. This compendium contains several useful articles on eighteenth-century regional maps of North America.

Sanchez-Saavedra, E. M. *A Description of the Country: Virginia's Cartographers and their Maps, 1607–1881.* Richmond: 1975. This slim volume and its accompanying maps offer a good survey of the cartography of Virginia.

Schwartz, Seymour, and Ralph Ehrenberg. *The Mapping of America*. New York: 1980. Chapters 6 and 7 of this ten-chapter book offer an excellent summary of mapping in eighteenth-century North America.

Snyder, Martin P. *City of Independence: Views of Philadelphia before 1800*. New York: 1975. This book, which contains many well-reproduced maps, is one of the few examples of the visually oriented city history. It contains much material useful for studying the development of the city in a spatial sense.

Stephenson, Richard W. *The Cartography of Northern Virginia: Facsimile Reproductions of Maps Dating from 1608 to 1915*. Fairfax Country: 1981. The first twenty plates in this publication refer to the period before 1800.

Tucker, Sara Jones, ed. *Indian Villages of the Illinois Country*. Springfield, Illinois: 1942. This volume, and its successors edited by Wayne Temple (1975) and Raymond Wood (1983), offer us a wide range of facsimiles of the printed and manuscript maps of the upper Mississippi valley; these volumes are available from the Illinois State Museum at Springfield.

Wheat, Carl I. *Mapping the Transmississippi West*. Vol. I. San Francisco: 1957. This volume, subtitled *The Spanish Entrada to the Louisiana Purchase, 1540–1804,* gives an exhaustive survey of the large-scale eighteenth-century maps of its area, and is particularly strong on the Spanish cartographers, who are often neglected. Unfortunately its reproductions of maps are sometimes rather fuzzy.

Plate 3.1. Detail from *A Map of the British Empire in America,* by Henry Popple (London, 1733).

A Detailed Anglo-French Map of the Interior, 1733

This detail comes from the very large map published in 1733 by Henry Popple, whose brother was an official in the Board of Trade and Plantations in London. It extends north to the edge of Hudson's Bay, south to Panama, east to Newfoundland, and west to the sources of the Mississippi River; by 1733, all this vast area was under reasonable cartographic control.

Our detail shows what is now southern Wisconsin and northern Illinois For this region, Popple would have had to rely exclusively on French sources, and especially on the work of Guillaume de l'Isle, who in 1718 published a *Carte de la Louisiane* that drew together the cartographic work of the previous half century. The map is accurate in its general delineation of the lake and rivers of this area; the "woods" and "mountains" are however almost entirely imaginary, serving a primarily decorative function.

Popple's engraving and typography are superb, and give the map a very special feel. Most of the names are French transliterations of the Indians place-names: "Ouiscongsing" for what is now Wisconsin, "Quicapous" for Kickapoos, and so forth. The portage at the future site of Chicago is well and carefully drawn, though there is as yet no settlement there. The only mission is the one of Saint François Xavier at what will become Green Bay. The forts on the Illinois and Chicago rivers are part of the great chain with which the French were hoping eventually to contain the anglophone colonies on the eastern seaboard.

Up on the Mississippi River, Popple mentions in English the existence of lead mines near what would become a major lead-mining district in the nineteenth century, centered on Galena, Illinois. He also notes an apparently mythical copper mine on the Illinois River. A map like this is an almost inexhaustible source of information about the state of the country when it was made, with the errors and omissions sometimes being as significant as the entries.

Source: Henry Popple, A Map of the British Empire in America, *ed. William P. Cumming and Helen Wallis (Lympne: England, 1972).*

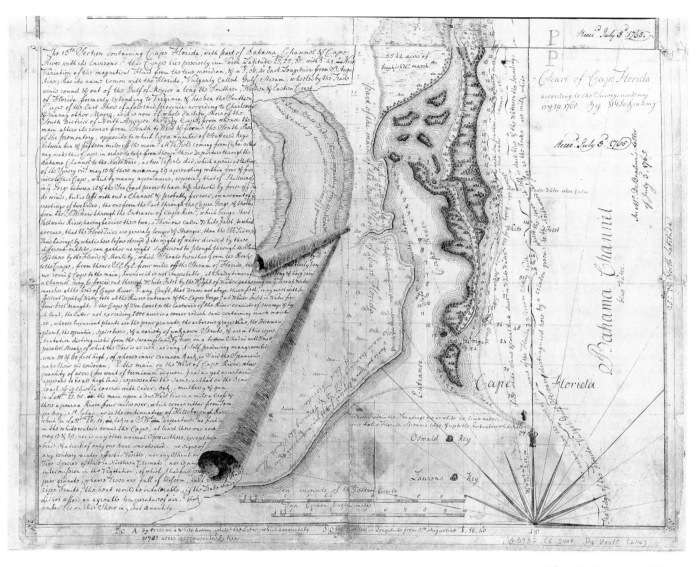

Plate 3.2. Manuscript "Chart of Cape Florida" by William Gerard de Brahm (1765). The Library of Congress.

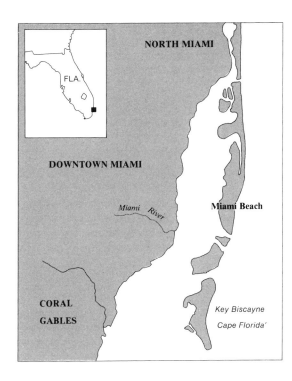

NORTH MIAMI

FLA.

DOWNTOWN MIAMI

Miami River

Miami Beach

CORAL
GABLES

Key Biscayne

Cape Florida'

Plate 3.2a. Modern map
covering the area of Plate 3.2.

An Image of the Virgin Florida Coastline, 1765

Samuel Holland and William De Brahm were among the cartographers whose work greatly expanded cartographic knowledge of North America during the 1760s. In fact, following their victory of 1763, the British appointed Holland surveyor general for the area north of the Potomac River, and De Brahm his counterpart for the area south of that. De Brahm was to concentrate on the area of what is now Florida, and in 1765 left Georgia, where he was provincial surveyor general of land, to begin his work.

The plate, showing a map completed in the first months of his incumbency, covers the area adjacent to the modern city of Miami in south Florida. Key Biscayne is the triangular-shaped island designated "Cape Florida." To the west (left), where downtown Miami now stands, the caption reads "here is good corn and

indigo land covered by gum, mulberry, oak and cidar [cedar]." On the land, De Brahm was thus concerned to indicate its possible usefulness; at sea, he was chiefly concerned to help navigators.

He has detailed soundings in fathoms, and does his best to offer verbal descriptions of the appearance of the water. Thus where the coastal waters meet the Bahama Channel, he has this legend: "The edge of the Florida, vulgo Gulf, Stream distinguished here by a celadon green, whilst the stream itself is dark blue and the waters on the soundings to the northward as far as the rocks are milk white". As the sketch map shows, this area of the coastline has greatly changed. But some features have a surprisingly degree of continuity: "Cape Florida" has largely retained its shape, and the Miami River is still recognizable, as is the general outline of the north-south spit. (Note that our sketch map has omitted some modern features such as bridges and causeways.) Indeed, De Brahm's map is not only a remarkably evocative document for historians, but also could be useful for ecologists and geologists in understanding the nature of the changes inflicted by man and by the sea on this once beautiful shoreline.

Sources: William P. Cumming, The Southeast in Early Maps *(Chapel Hill: 1962); Louis De Vorsey, ed.,* De Brahm's Report of the General Survey in the Southern District of North America *(Columbia: 1971); and Roland Chardon, "A Best-Fit Evaluation of De Brahm's 1770 Chart of N. Biscayne Bay,"* The American Cartographer *IX (1982), pp 47–87.*

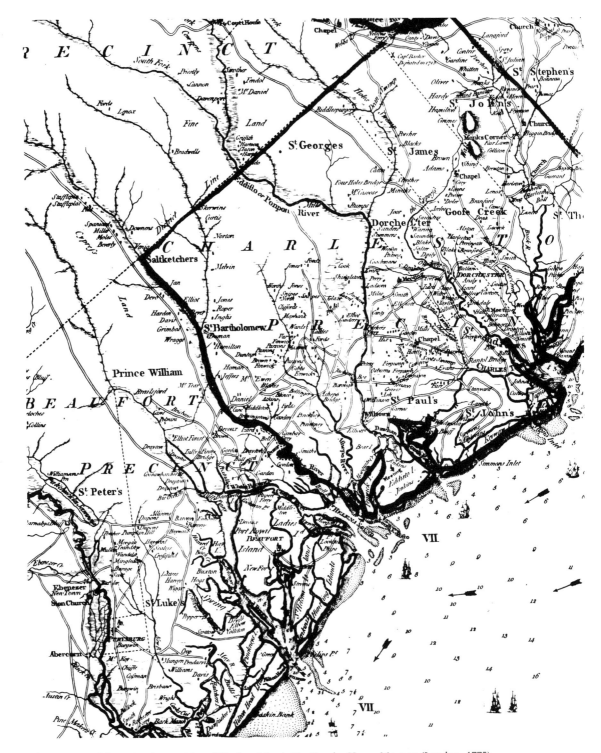

Plate 3.3. Detail from *An Accurate Map of North and South Carolina*, by Henry Mouzon (London, 1775).

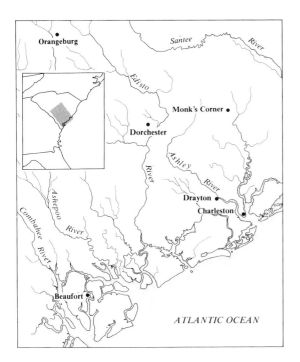

Plate 3.3a. Modern map
covering the area of Plate 3.3.

Details of the Early Settlement of South Carolina, 1773

This map, like most of those published at this time, drew heavily on previous work. Henry Mouzon of South Carolina was, however, himself an active surveyor who made many corrections to the existing work, and this map of his gave the best possible image of the Carolinas on the eve of the Revolutionary War. George Washington had a copy of this map, folded and cloth-backed for use in the field; it is now in the library of the American Geographical Society, at Milwaukee.

A short examination of the map's hydrographic elements, in comparison with those of the sketch from a modern map (Plate 3.3a), will show how accurate Mouzon was in drawing the main features. The Combahee, Ashepoo, Edisto, Ashley, Cooper, and Santee rivers are all very correctly shown, and cities like Orangeburg, Beaufort, Dorchester, and Charles Town (now Charleston) are well plotted. There is also some attempt to show soundings off the coast (in fathoms, six feet each).

This part of Mouzon's map shows much of the "Charlestown Precinct," an early name for an administrative subdivision. It also shows the boundaries of many of the parishes: "St. George's," "St. James," "St. Stephen's," and so forth, and takes care to indicate the various places of worship: "church," "chapel," and [Quaker] "meeting," with a symbol. Some of these parish names have become place-names, such as "St. George," and some distinctive town names, including "Monks Corner," have survived.

There are many levels at which one can interrogate a map of this kind. We have said nothing of the road patterns and their possible relation to Indian trails, or of the individual property names, though "Drayton" has been marked on the sketch map to indicate the site of one of the most famous estates, from which a magnificent house survives. The map is in fact so dense with information about late eighteenth-century South Carolina that it would merit inch-by-inch examination, with continuous comparison with a suitable modern map.

Sources: Thomas Jefferys, The American Atlas *(London: 1776, facsimile Amsterdam, 1974, ed. Walter Ristow); and* North America at the Time of the Revolution, *part III, ed. William P. Cumming and Douglas Marshall (Lympne: England, 1975).*

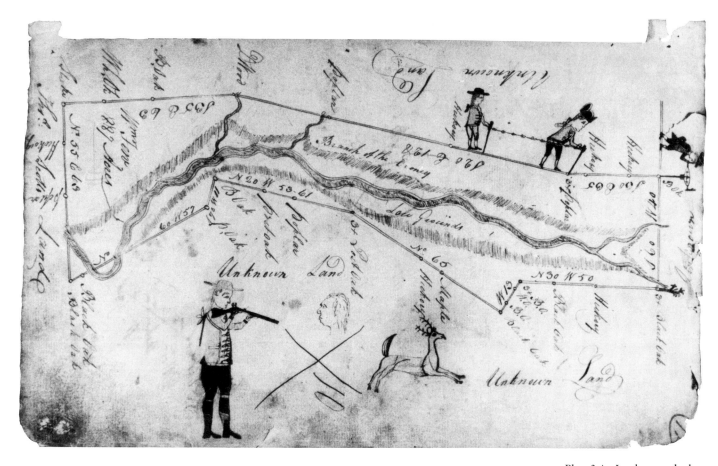

Plate 3.4. Land-grant plat by William Few, c. 1780. The Georgia State Department of Archives and History.

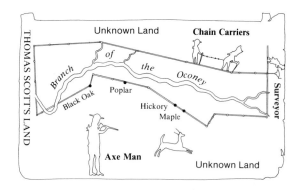

Plate 3.4a. Diagram of Plate 3.4.

A Survey Party at Work on the Ground

This plat is typical of many preserved in archives all over the country, basically showing the geometrical outline of a landholding. It was constructed in a series of measured straight lines, proceeding from one landmark to another; these landmarks are shown on the plat by small circles. When these circles were joined together on paper, they indicated the "metes and bounds," or limits, of the property. What makes this little map remarkable are the identity of the owner and the incidental decoration.

William Few was one of the two Georgia patriots who helped frame the Constitution of the United States in 1787. The surveyor or someone associated with him sketched the surveying party as they carried out their task. Near the right-hand margin the surveyor himself is seen, sighting on a "black oak" corner tree through the dovetail sights mounted on his large surveyor's compass. Coming along behind the surveyor are two chain carriers, measuring the metes and bounds of the area with a Gunter's Chain; this was made up of 100 links and equaled sixty-six feet. The man below with the flintlock is no doubt the axe man, whose task was to blaze distinctive marks in the trees marking the surveyed lines.

The attire of the survey party is revealing. The surveyor himself is wearing the breeches, frock coat, and planter's broad-brimmed hat of a well-to-do citizen of the day. The chain carriers and axe man, on the other hand, seem to be wearing bits and pieces of Revolutionary War uniforms, and are probably recently discharged veterans. We have here an intimate glimpse of the process of land surveying and the type of people who were extending the settlement frontier in eighteenth-century North America.

Notice that William Few's land was surrounded on three sides by what is termed "unknown land." In the parlance of the times this meant ungranted land, available for acquisition. Few's neighbor lower down on the "branch of the Oconey," as the creek is designated, was Thomas Scott. The surveyor was careful to show something of the topography of Few's grant by a sort of hachuring to indicate a low terrace rising above the "Low Grounds" that form the stream's flood plain.

A close examination of the plat will reveal that hickory, black oak, poplar, white oak, dogwood, maple, and post oak were chosen as witness trees along the boundaries of this parcel of land. The list is suggestive, but should not be taken as a complete count, for surveyors usually chose as their witness trees the more enduring hardwoods, rather than pines or other conifers. Many plats bear evidence of prior Indian use of the land, such as Indian mounds, village sites, paths, or old fields. On the Few plat only the river name "Oconey" [Oconee] serves to remind us of the prior occupants of this frontier landscape in northeast Georgia.

Source: Louis De Vorsey, "Early Maps as a Source in the Reconstruction of Southern Indian landscapes," in Red, White and Black: Symposium on Indians in the Old South, *ed. Charles M. Hudson, (Athens: 1971).*

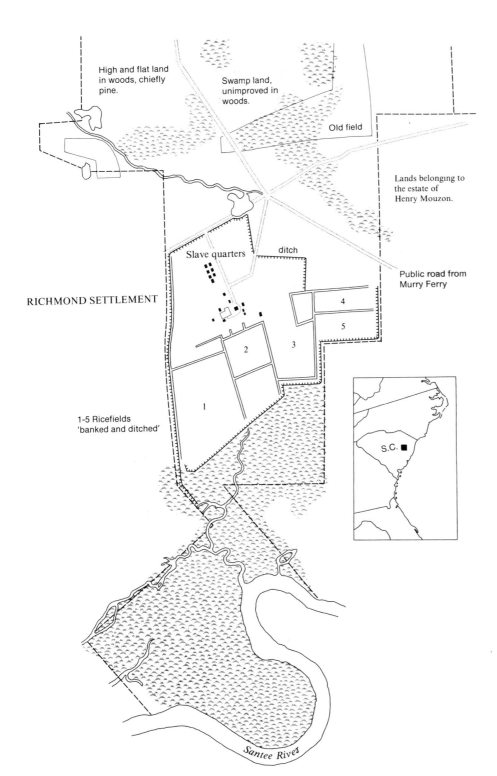

High and flat land in woods, chiefly pine.

Swamp land, unimproved in woods.

Old field

Lands belonging to the estate of Henry Mouzon.

Slave quarters

ditch

Public road from Murry Ferry

RICHMOND SETTLEMENT

4

5

3

2

1

1-5 Ricefields 'banked and ditched'

S.C.

Santee River

Plate 3.5. Redrawing of Joseph Purcell's "A Plan of Richmond Plantation," 1789–91. South Carolina Historical Society, Charleston.

The Detailed Layout of a South Carolina Rice Plantation, c. 1790

Many estates were carefully delineated in England, Wales, and Scotland, and also in the West Indies, from the seventeenth century onward, and these plans are of great interest to the historian. Relatively few seem to have survived for the United States, but for the prosperous estates of the late eighteenth century they are fairly abundant in the archives of South Carolina. The great problem in using them is that they are often too stained and battered to be legible on photographs. So it is often necessary to do as we have done here, and make a modern map from them, following the ideas put forward in the article by Wiberly (see the source at the end of this plate description).

One of the most prolific estate cartographers was Joseph Purcell. Our sketch shows a detail from one of his plans, now preserved at the Historical Society of South Carolina in Charleston. Its full title is "A plan of Richmond plantation . . . belonging to Theodore Gaillard Esquire, situated on the north side of Santee River, Prince Frederick Parish, Georgetown District," and it was drawn between 1789 and 1791. Richmond Plantation lay in the topmost right-hand corner of our Plate 3.3, where indeed "Galliard's Island" may be seen.

The sketch shows the Santee River running along the southern edge of the holding, with extensive "swamp lands" to the north of it. Then we come to the heart of the plantation, the rice fields, close to which are the dwellings both of the master and of the slaves (the latter the seven huts in two rows). North again we come in the west to the high lands where pines thrive, and in the east to more swamps, giving onto the lands of our friend Henry Mouzon. Here, too, passes the road to Murry Ferry, visible at the top right on Plate 3.3.

Our drawing has had to be somewhat simplified; in particular, it has not been possible to reproduce the very exact field measurements and specifications, or the many mentions of tree types along the various boundaries. In fact, as we should expect, pines predominate in the northern section, while along the southern edges are cypress, white oak, hickory, maple, gum, and water oak. A map like this really needs to be read in close conjunction with the written records of the estate, for we have here a striking representation of its spatial arrangement, and lack only the progression of change over time—which might be found in other types of estate record—to bring Richmond Plantation back to life.

Source: Stephen E. Wiberly, Jr., "Editing Maps: A Method for Historical Cartography," Journal of Interdisciplinary History X (1980), pp 499–510.

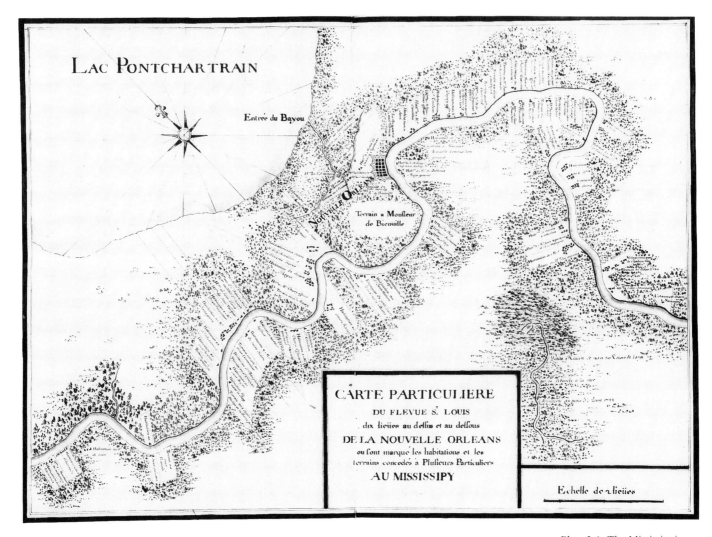

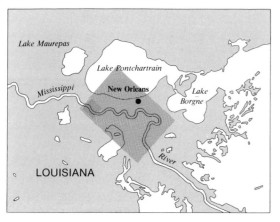

Plate 3.6. The Mississippi River in the neighborhood of New Orleans; manuscript map from the "Cartes Marines," c. 1720.

Plate 3.6a. Area shown on plates 3.6 and 3.6b superimposed on modern map of the delta.

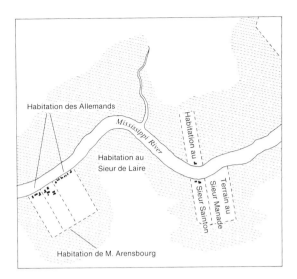

Habitation des Allemands

Mississippi River

Habitation au Sieur de Laire

Habitation au

Terrain au Sieur Manadé

Sieur Sainton

Habitation de M. Arensbourg

Plate 3.6b. Detail of area shown in Plate 3.6.

The Early French Establishment on the Lower Mississippi

One step down in scale from plats and estate plans are maps that show a number of estates. They are rare, but often very revealing. One such map is this early eighteenth-century image of the Mississippi River above and below New Orleans. The city had been founded about five years before this map was drawn, in 1723, and at the same time lots had been taken up along the river. They are in the characteristically French pattern known as *long lots,* whereby each landholder had access to the river, the lots running away as long narrow strips at right angles to it.

From the sketch map (Plate 3.6a), and still more from the satellite image (see Color Plates 1 and 2), it is plain that our map is accurate in its delineation of the river's windings. New Orleans is shown with the town laid out on a grid pattern, and protected by a bastioned fortification. The names of the landholders are written in their lots just as the lots run, which is

why the comparison between the map and the satellite image is so close.

A map like this (Plate 3.6b) holds in compressed form an immense amount of information. Let us go to the lower left corner (see the box on the location map), and analyze the first half-dozen entries. The first and third read "Habitation des Allemands," referring to the 300 or so Germans who had been recruited by John Law and were established here in 1722. They suffered greatly from fever and floods, but eventually succeeded in making a solid establishment. Their leader at this time was Charles-Frédéric d'Arensbourg, a Swedish captain; his name appears on the second plot. In between the first three plots and the next ones is the "Habitation au sieur de Laire"; this is the land of Michel Delaire, an early entrepreneur whose venture had already failed. Downstream we come to the "Habitation au sieur Sainton." Pierre Sainton was the son of a lawyer from Châtellerault in France who was trying to establish himself here. Finally, we come to the "Terrain au sieur Manadé," land of Pierre de Manadé, a surgeon who was adapting well to the new life. He raised cattle, perhaps on the very *terrain*—a word indicating less permanent settlement than *habitation*—and had taken an Indian to be his wife.

We have looked only at one small corner of this fascinating map, and yet are beginning to have some idea of the nature of the settlement. It would be an excellent teaching exercise, or even a good research project, to carefully edit a map like this, using teams if necessary, so as to translate the spatial information into an historical account by linking it with the written sources.

Source: Marcel Giraud, Histoire de la Louisiane Française, *4 vols. (Paris: 1953–1974; vol. 1, 1974).*

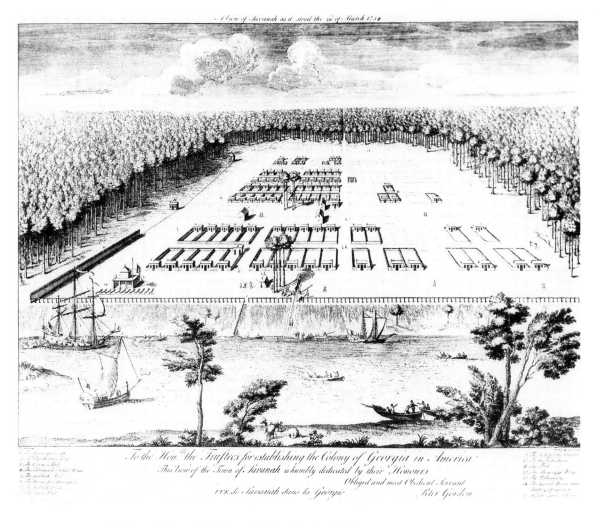

To the Hon.ᵇˡᵉ the Truſtees for eſtabliſhing the Colony of Georgia in America
This View of the Town of Savanah is humbly dedicated by their Honours
Obliged and most Obedient Servant
VUE de Savannah dans la Georgie.
Peter Gordon

Plate 3.7. *View of the Town of Savannah,* by Peter Gordon (London, 1734).

An English City in the Wilderness

Original town plans and surveys are often of great value and interest to students, who enjoy picking out familiar sites or identifying particular stages in the growth and change of their city. Plans can often, too, reveal clues to the goals and ambitions of town founders. One of the most interesting early plans of this kind is the one of Savannah, Georgia, prepared in 1734.

This engraving is deemed to be, in its original state, one of the rarest American urban prints. Thanks to photo-facsimile technology it is frequently reproduced and has become the icon of New World urban planning.

Although his name figures prominently in the dedication of this engraving, Peter Gordon was not its author. He was an unem-

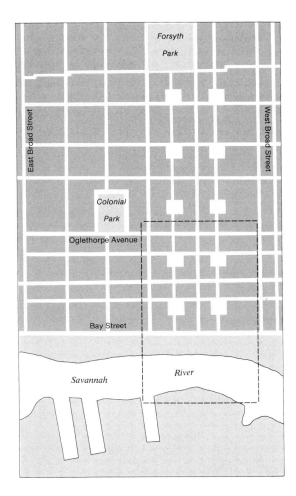

Plate 3.7a. Area shown on Plate 3.7 (dashed line) superimposed on modern plan of Savannah.

the form of a "compleat drawing" of the oblique perspective view which served as the model for a copperplate engraving executed by P. Fourdrinier. Oglethorpe himself left Savannah on March 29, 1734, and when he arrived in London he had the buildings and facilities completed between November and the end of March added to the drawing and copperplate to bring it to the state we see here.

However, the engraving is, in a sense, incomplete in that it shows only four of Savannah's original six wards, laid out by Oglethorpe and South Carolinian Colonel William Bull with town surveyor Nobel Jones, on the level sandy bluff overlooking the waters of the Savannah River. Two additional wards with their open central squares were surveyed to the left or east of the four shown in the view. Rather than being placed outside the town as it might appear, "The Parsonage House" (no. 12) was centrally located on the public or "Trust" lot which was just across present day Drayton Street from "The Lott for the Church" (no. 9). The visually impressive "Pallisadoes" (no. 13) were found to be unnecessary after only a few days of construction when Oglethorpe realized that the local Indians posed no threat. Rather, chief Tomo Chichi and his band were organized to form "two companies . . . of Forty very Clever Men. Their pay is one Bushell of corn pr. month for each man while we employ them in War or hunting, a Gun at their first listing and a Blanket p. ann."

The engraving has many other features of interest: the communal or "public" mill and oven, the crane used for hauling goods up the bluff from the ships, the ships themselves, and perhaps most evocative of all, the surrounding stands of huge pine trees, reminding us that these were indeed cities in the wilderness.

Source: George F. Jones, "Peter Gordon's (?) Plan of Savannah," *The Georgia Historical Quarterly LXX* (1986), pp. 97–101; Rodney M. Baine and Louis De Vorsey, Jr., "The Provenance and Historical Accuracy of 'A View of Savannah as it stood the 29th of March, 1734,'" *The Georgia Historical Quarterly* (in press).

ployed London upholsterer who had been chosen as one of the original contingent who sailed with James Edward Oglethorpe and founded the town of Savannah and Colony of Georgia in 1733. In November of that year Gordon was forced to return to London to undergo an operation. Oglethorpe took advantage of the opportunity and entrusted him with a plat or map of the town and sketches of the buildings being constructed in Savannah. A London carpenter-artist compiled and rendered this information in

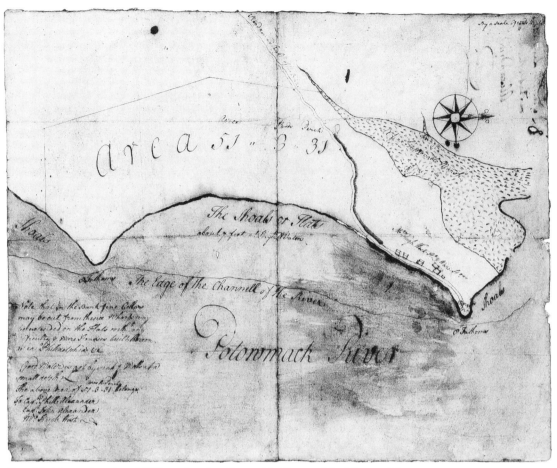

Plate 3.8. Manuscript plan of 1748, attributed to George Washington. The Library of Congress.

Washington's Plans of the Alexandria Site

Plate 3.8 shows a manuscript plan thought to have been drawn by George Washington in 1748 when he was seventeen. With manuscript in hand, the terms of the act that authorized the foundation of the town of Alexandria are easy to follow: "beginning at the first branch above the warehouses, and extending down the meanders of the River Potomac to a point called Middle Point and thence down the river ten poles."

The warehouses mentioned are shown on the map, and identified as "Mr. Hugh West, Ho. and Ware Hoss.," built along the "Road round Hd. of the Crk." The town plan, Plate 3.9, was probably prepared the following year, prior to the organization of the municipal government, which was instituted on July 13, 1749. The list of names on the right of the plan, headed "No." and "Proprietors' Names," indicates who purchased what lots and the price they paid (in pistoles) in the first land sale, held in July 1749. Like the town site plan, the town plan is a pen-and-ink manuscript in the Library of Congress; both are attributed to George Washington, in spite of their radically differing styles.

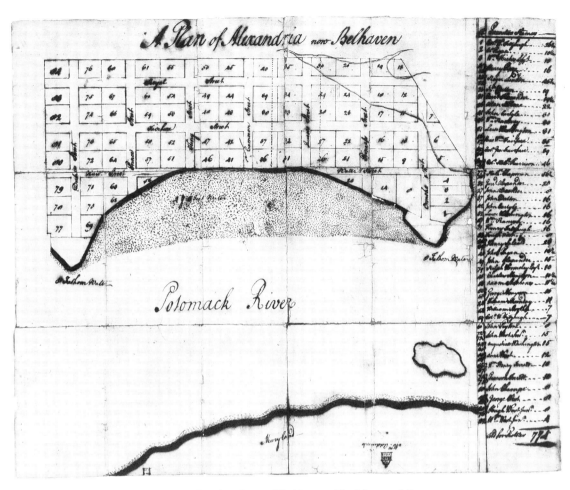

Plate 3.9. Manuscript plan of 1749, attributed to George Washington. The Library of Congress.

Both the 1748 map and the 1749 plan emphasize Alexandria's relationship to the Potomac River, and particularly the deep-water navigation channel of the stream. Just off the two points, which came to be named West and Lumley, Washington has marked the depth of eight fathoms or forty-eight feet; this was more than enough for any vessel of that day to find safe steerage. The broad cove between the points was described as "The Shoals or Flats about 7 feet at high water." Another gloss on the map states: "Notes that in the Bank fine cellars may be cut; from thence wharfs may be extended on the Flats without any difficulty & warehouses built thereon as in Philadelphia."

The early history of Alexandria shows that wharves were in fact built out to the navigation channel in short order, and the town was extended across the former flats, with a new street named Union eventually connecting Point West and Lumley. Much of the attractive historic district of present-day Alexandria is actually built on land reclaimed from the tidal shallows, just as Washington predicted on his eighteenth-century map.

Source: John Reps, The Making of Urban America (Princeton: 1965).

CHAPTER FOUR

Maps of the Township and Range System

Ronald E. Grim

After the Revolutionary War, the new government of the United States faced the problem of how to control the newly acquired land outside the thirteen original states. There was general agreement that this vast area could not be allowed to develop piecemeal, as had the eastern seaboard, and some form of rectangular survey seemed to be the best solution. But what should be the size of the units? Thomas Jefferson pressed for units ten miles square, in accordance with the decimalizing tendencies of the time, but in the end the Land Ordinance of 1785 fixed on "townships" six miles square as the basic unit.

The land management system that emerged from this ordinance not only had a profound effect on the physical appearance of much of the western country, but also produced a massive set of cartographic and other records documenting the initial survey and disposal of the public lands. These records have been used primarily by professional surveyors, cartographers, and geographers, but they are also an essential source for historians studying the settlement process and local history of the thirty public land states (all the states except the original thirteen, and Kentucky, Tennessee, Vermont, Maine, West Virginia, Texas, and Hawaii; see Plate 4.1).

Following the provisions of the Land Ordinance of 1785, the first surveys were carried out between 1785 and 1788 in a district of Ohio known as the *Seven Ranges* (Plate 4.3). These were the work of Thomas Hutchins, Geographer of the United States. As the years went by the survey gradually worked westwards, with the surveyor general of each new territory contracting with deputy surveyors to get the work done. In 1812, the General Land Office was established to oversee the land transactions that resulted from the progressive expansion of the public domain, and after 1836 the surveyors general were incorporated into that agency, which in 1946 became part of the Bureau of Land Management.

The cadastral system that was adopted is based on a rectangular grid, a radical departure from the irregular pattern of metes and bounds surveys (see, for example, Plate 3.4) used in the original thirteen states. The basic unit is the township, measuring six miles square, and numbered north or south and east or west from a succession of numbered or named principal meridians and base lines. Taking Illinois as an example of how this system works, note on Plate 4.1 that the state has a meridian, the "3rd Principal Meridian," running down its center, and an east-west "base line" cutting across its southern third. On Plate 4.2, this cross is shown on the state map, and a detail is also shown from the upper right-hand area of the cross. Here is a checkerboard of townships, numbered north from the crossing point in *townships* (a confusingly ambiguous term) and east from it in *ranges*. As the plate shows, each township is then divided into thirty-six sections, each measuring 1 square mile or 640 acres, and numbered from one to

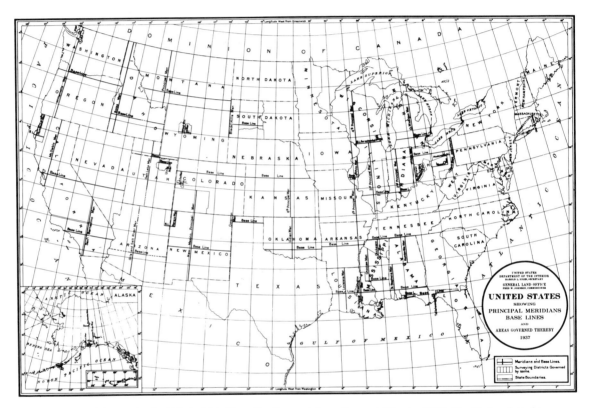

Plate 4.1. *United States, showing Principal Meridians, Baselines and Areas Governed Thereby,* 1937. The National Archives.

Map to show area covered by the township and range system

This official map of 1937 shows the principle baselines and meridians governing the system. Note that a large part of the east is not covered; here the very landscape differs sharply from the area moulded by the Land Ordinance's grid. Just to the west of the word "Pennsylvania" is the area on the Ohio River where the original Seven Ranges (Plate 4.3) were laid out; note also the cruciform shape of the meridian and baseline in Illinois (Plate 4.2).

thirty-six, and these sections can be broken down into acreages.

The basic cartographic records produced as a result of the township surveys were the field notes and three manuscript copies of each township plat. The original field notes were recorded in the field, and then returned to the surveyor general's office, where three manuscript plats were prepared for each township, as well as a transcript of the field notes. These plats show not only the surveyed lines delineating the township, but also some hydrography, vegetation, natural resources, and cultural features such as trails, roads, and Indian and pioneer settlements. One copy of the plat and the original field notes were retained by the surveyor general and eventually became the state copy; the second copy and the transcribed field notes went to the General Land Office in Washington, D.C., and the third copy was used in the local land office, where it was annotated with references to specific land transactions. The present location of these documents for all of the relevant states is set out in the Appendix.

Since the plats and field notes are primarily surveying documents, their basic value is for surveyors and lawyers who need to reconstruct township and section lines and property boundaries. In addition, however, the plats in conjunction with the field notes have much potential in reconstructing the physical and cultural landscapes during the initial phase of an area's settlement history, which often represented the beginning of intensive agricultural and urban use of the area; some secondary works using them in this way are also listed in the bibliography at the end of this chapter.

Based on the principle of survey prior to the sale of land, most areas were surveyed prior to settlement or early in the settlement process. However, the plats do not provide a comprehensive topographic survey of an area, though as time went by they tended to show more detail of this kind. In fact, consistency of information may be a concern in the reconstruction of landscapes for large areas. As the survey system developed, standardized instructions were issued and eventually published in the form of a manual of instructions, the *Instructions to the Surveyors General of Public Lands of the United States for those Surveying Districts Establishes in and since the year 1850* (Washington, D.C., 1855). Before this time there was a good deal of variation in the way the many deputy surveyors approached their task.

Even after 1855, township surveying remained an arduous and somewhat approximate process, sometimes requiring twenty or thirty years to complete all the surveys in a state. As there were many deputy surveyors, possibilities of error, omission, personal bias, and even misrepresentation abounded. From start to finish, then, the process of surveying was far from standard or uniformly regulated, so that we should be on the alert for possible variations in the quality of the material recorded.

While the original and duplicate plats are most useful for the reconstruction of past landscapes, the triplicate plats (with ownership notes) are helpful in recreating original land-ownership. However, these plats must be used in conjunction with other sources. Of primary importance are tract books and land entry papers that are part of the records of the former General Land Office (Record Group 49) in the General Archives Division of the National Archives. They contain descriptive entries of each land transaction by township and section, indicating name of purchaser, type of transaction, date of purchase and patent, and a listing of corresponding land entry files. The latter listing serves as an index to the land entry papers or case files, which include various papers such as the applications, receipts, warrants, or in the case of homestead entries, final proofs of settlement that were accumulated during the process of recording the various land transactions. By

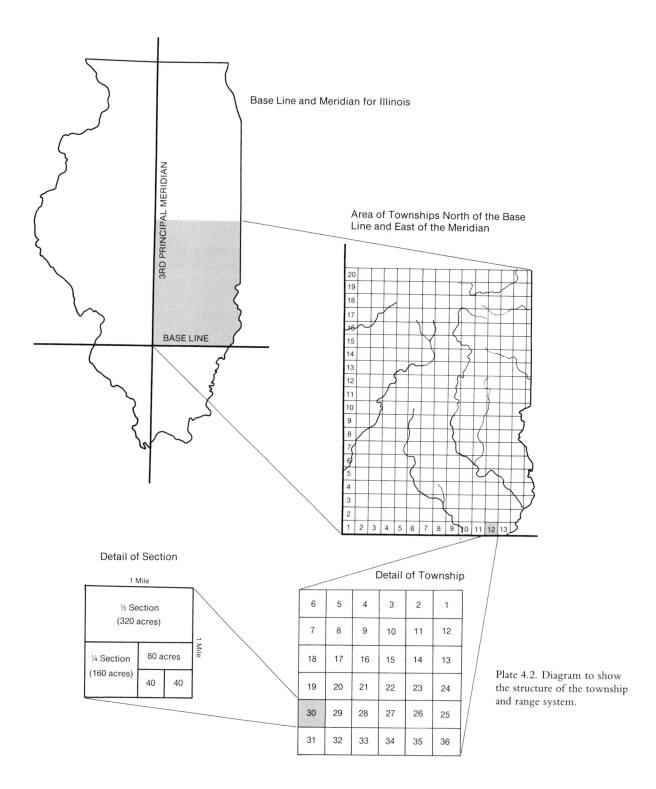

Base Line and Meridian for Illinois

3RD PRINCIPAL MERIDIAN

BASE LINE

Area of Townships North of the Base Line and East of the Meridian

Detail of Section

1 Mile

½ Section
(320 acres)

1 Mile

¼ Section
(160 acres)

80 acres

40 | 40

Detail of Township

6	5	4	3	2	1
7	8	9	10	11	12
18	17	16	15	14	13
19	20	21	22	23	24
30	29	28	27	26	25
31	32	33	34	35	36

Plate 4.2. Diagram to show the structure of the township and range system.

correlating these three sets of records it is possible to plot cartographically the original land-ownership patterns for a given township. Such a compilation can provide a base for comparing subsequent patterns as depicted in the commercially published county maps and atlases described in Chapter 8.

Although the township plats are the basic cartographic record resulting from the General Land Office surveys, they were not the only maps produced. Among the cartographic records of the General Land Office in the National Archives are several large series pertaining to special surveys: private land claims, townsites, mineral surveys, state boundary surveys, and railroad rights-of-way. In addition, there is a general map collection, the Old Map File, which consists of manuscript and annotated maps showing the extent of surveys in individual states, surveying activities in individual land districts, regional maps depicting private land claims, and surveys of special reserves such as military or Indian reservations. There are also collections of General Land Office published maps that contain maps of the entire United States and of individual states showing the progress of surveys at various intervals. When used in conjunction with the township plats, the special surveys provide additional background about the more unusual survey and settlement features in an individual locality, while the more general maps help place the individual locality in the context of a broader regional setting.

In using this material, we ought to bear in mind that it was the product of a unique cartographic situation. Never before had there been a great mass of humanity, poised to occupy, acre by acre, so huge an area; this was not a conquest by military force alone, but also by mass settlement. It is hard to imagine what sort of chaos might have ensued had the Land Ordinance not given an orderly framework to the process of settlement. Of course, this process was marred by various forms of double-dealing and corruption, or even by the simple fact that many veterans, for instance, received land that they did not retain, but straightaway sold to land speculators. It was also marred in places by inaccurate land measurement, which in time gave rise to disputes. Still, by and large the enterprise was astonishingly successful. It gave the lands covered by the survey a Mondrian-like square quality that they retain to the present day. Particularly when going by air, the traveler across the United States is acutely aware of passing from the disorderly east into what might be called the landscape of the Enlightenment, with a serious attempt to impose the rational grid upon a sometimes unruly nature.

At times, of course, the grid was a nuisance; roads tended to develop along the section lines, so that it was often—and still may be—impossible to reach a point on a distant diagonal except by going north-south and then east-west. Also, for some people the rectangular appearance of the countryside is monotonous, though in fact the rectangles are often broken by offsets necessary to take into account the spherical nature of the earth. In the end, though, the grid was not only essential to the accurate mapping of a huge area, but also made relatively easy such huge projects as the financing of education (section 16 was usually set aside for school buildings) or the construction of public works like the Illinois and Michigan Canal, which lay at the origins of Chicago. If the country had not been divided up into manageable sections, it would have been almost impossible to visualize the financing of projects of this nature, which had to pay for themselves by land sales. In conjunction with the Northwest Ordinance of 1787, the Land Ordinance of 1785 went far to determine the economic structure, as well as the physical appearance, of a huge area of the United States.

SOURCES FOR GENERAL LAND OFFICE MATERIAL

WORKS DEALING WITH THE HISTORY OF THE TOWNSHIP AND RANGE SYSTEM

Cazier, Lola. *Surveys and Surveyors of the Public Domain, 1785–1975*. Washington, D.C.: 1976. Contains interesting illustrations of surveyors at work.

Ehrenberg, Ralph E., ed. *Pattern and Process: Research in Historical Geography*. Washington, D.C.: 1975. This volume contains papers prepared for the Conference on the National Archives and Research in Historical Geography held in 1971. For our purposes here, note particularly Hildegard Binder Johnson. "The United States Land Survey as a Principle of Order"; and William D. Pattison. "Reflections on the American Rectangular Land Survey System."

Ernst, Joseph. *With Compass and Chain: Federal Land Surveyors in the Old Northwest, 1785–1816*. New York: 1979. A good summary of the operation of the township survey system during its first three decades.

Gates, Paul W. *History of Public Land Law Development*. Washington, D.C.: 1968. A substantial work, particularly strong on the legal aspects of the process of land division.

Johnson, Hildegard Binder. *Order upon the Land: the U.S. Rectangular Land Survey and the Upper Mississippi Country*. New York: 1976. An interesting summary of the origins of the system, and then its implementation and impact in a specific area.

McEntyre, John G. *Land Survey Systems*. New York: 1978. This book is "intended to serve both as a textbook for students in land surveying programs and as a general reference book for the practitioner," but it also contains much technical information of interest to historians.

Pattison, William D. *Beginnings of the American Rectangular Land Survey System, 1784–1800*. Chicago: 1957. A meticulous and instructive examination of the origins of the survey, with an extensive bibliography.

Pattison, William D. *A Partial Answer to the Question, Who has Exploited the Public Land Survey Plats and Field Notes for their Descriptive Content?* (Typescript at the Newberry Library, Chicago, n.p., n.d.). A very interesting and suggestive summary of the way in which these records have been, and could be, used.

Rohrborough, Malcolm J. *The Land Office Business: The Settlement and Administration of American Public Lands, 1789–1837*. New York: 1968. A narrative history relating the process of land division to the wider political issues of the day.

Rooney, John F.; Wilbur Zelinsky; and Dean R. Louder, eds. *This Remarkable Continent: an Atlas of United States and Canadian Society and Cultures*. College Station, Texas: 1982. The chapter by Terry G. Jordan, "Division of the Land," contains many striking maps to summarize developments in the various survey systems found in North America.

Stewart, Lowell O. *Public Land Surveys: History, Instructions, Methods*. Minneapolis: 1976: Originally published in 1935, this little book remains one of the best descriptions of the nuts and bolts of the survey process.

Thrower, Norman J. W. *Original Survey and Land Subdivision: A Comparative Study of the Form and Effect of Contrasting Cadastral Surveys*. Chicago: 1966. This stimulating little book investigates not only the process of survey, but also some of its wider consequences.

White, C. Albert. *A History of the Rectangular Survey System*. Washington, D.C.: 1983. A compendious history of the surveys, whose great merit is to reproduce many of the original documents regulating the process.

SECONDARY WORKS MAKING PARTICULARLY EFFECTIVE USE OF THE MATERIAL GENERATED BY THE TOWNSHIP AND RANGE SURVEYS

Birch, Brian B. "The Environment and Settlement of the Prairie-Woodland Transition Belt—a Case Study of Edwards County, Illinois." *Southampton Research Series in Geography* 6 (1971), pp. 3–31.

Durbin, Richard D., and Elizabeth Durbin. "Wisconsin's Old Military Road: Its Genesis and Construction." *Wisconsin Magazine of History* 68 (1984), pp. 2–42.

Franzwa, Gregory M. *Maps of the Oregon Trail*. Gerald, Missouri: 1982.

Hanson, Philip. "The Presettlement Vegetation of

the Plain of the Glacial Lake Chicago in Cook County, Illinois." *Ohio Biological Survey Biological Notes* 15, pp. 159–64.

Jordan, Terry G. "Between the Forest and the Prairie." *Agricultural History* 38 (1964), pp. 205–16.

Knox, James C. "Human Impacts on Wisconsin Stream Channels." *Annals of the Association of American Geographers* 67 (1977), pp. 323–42.

Mansberger, Floyd. "Initial Field Location in Illinois." *Agricultural History* 57 (1983), pp. 289–96.

McManis, Douglas R. *The Initial Evaluation and Utilization of the Illinois Prairies, 1815–1840.* Chicago: 1964.

Moran, Robbin C. "Presettlement Vegetation of Lake County, Illinois." In *Proceedings of the Fifth Midwest Prairie Conference,* (1978), pp. 12–18.

Schroeder, Walter A. *Pre-Settlement Prairie of Missouri.* Missouri: 1981.

Sears, Paul Bigelow. "The Native Vegetation of Ohio." *Ohio Journal of Science* 25 (1925), pp. 139–49 et seq.

Stocking, Hobart E. *The Road to Santa Fe.* New York: 1971.

Walters, William D. "Early Mill Location in Northern Illinois." *Bulletin of the Illinois Geographical Society* 25 (1983), pp. 3–12.

LOCATION OF GENERAL LAND OFFICE TOWNSHIP PLATS AND FIELD NOTES

The Appendix lists the addresses and telephone numbers of institutions holding one or more sets of General Land Office plats and/or field notes. It was compiled by telephone conversations from December 1984 to April 1985, and not by personal examinations; consequently, some errors are likely in the scope of collections attributed to some institutions.

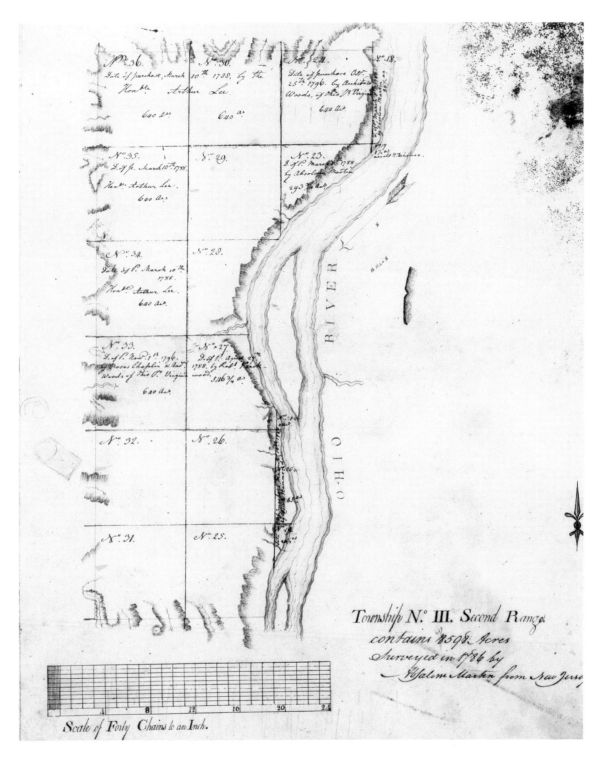

Plate 4.3. Plat of Township 3, Range 2, in the "Seven Ranges" of Ohio, surveyed in 1786 by Absolom Martin. The National Archives.

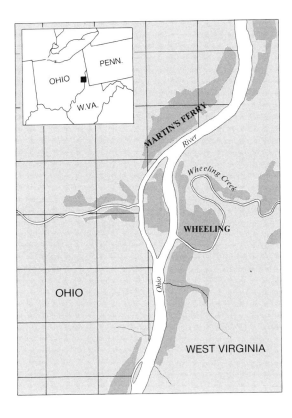

Plate 4.3a. Modern map covering the area of Plate 4.3.

The Beginning of the System

The first surveys resulting from the Land Ordinance of 1785 were conducted in southeastern Ohio, in an area now referred to as the *Seven Ranges*. This plat of Township 3 in Range 2 was surveyed in 1786 by Absalom Martin of New Jersey, and was perhaps drawn by Surveyor General Thomas Hutchins. The area covered (see Plate 4.3a) was on the west bank of the Ohio River, opposite the present site of Wheeling, West Virginia. Here there was intense land speculation during the late eighteenth century, in some of which the surveyors were personally concerned.

This plat depicts with considerable accuracy the topography and streams along the exterior boundaries and on the western bank of

the Ohio River. But it does not record similar information along the section lines, and contains virtually no information concerning cultural features. The numbering of sections in this incomplete (fractional) township is not in conformity with later standards for it begins at the top right with section 18, by the Ohio River, and then works westwards, from bottom to top. Note that the scale is in *chains,* deriving from the 22-yard chain like the one carried by the surveyor's aides in Plate 3.4. The chain has generally passed out of use, except in the formerly British West Indies, but it remains the length of a cricket pitch.

On the plat there is a note that sections 18 and 23, "prime sites" along the Ohio River, were in fact bought by Absalom Martin, the surveyor; on these lands he established the town of Martins Ferry in 1787. As the survey moved westwards, the information included on the plats tended to become fuller, and the layout of the township became standardized into the six-by-six mile square. The scale, though, remained the same, at one inch to half a mile, and this proved adequate for conveying a good deal of information.

Source: William D. Pattison, Beginnings of the American Rectangular Land Survey System, 1784–1800 *(Chicago: 1957).*

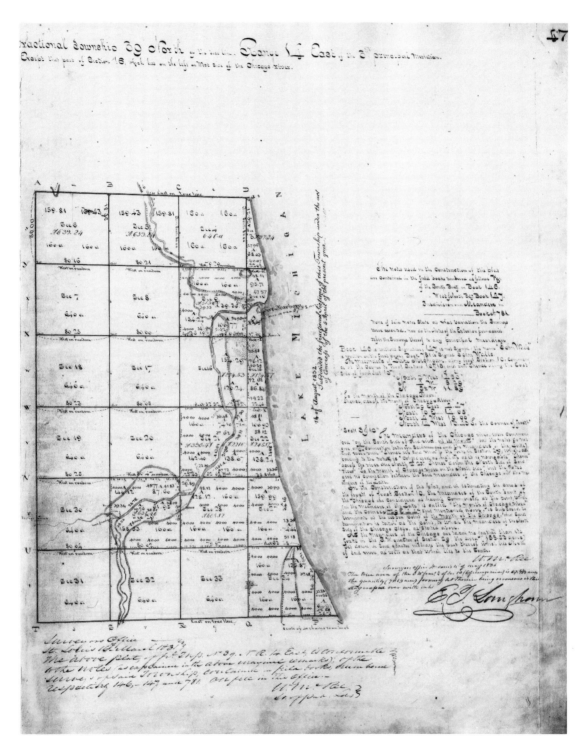

Plate 4.4. Plat of the Fractional Township at the mouth of the Chicago River, 1831.
Illinois State Archives, Springfield, Illinois.

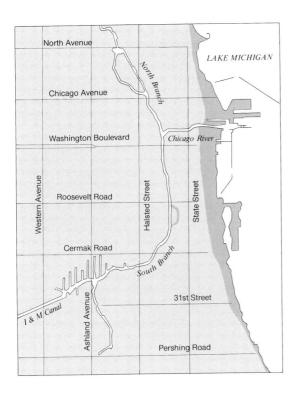

Plate 4.4a. Modern map
covering the area of Plate 4.4.

A City in the Making

This plat, of the area where Chicago would soon rise, shows the region as it was when the surveyors first reached it. There is no sign of settlement at the mouth of the river, but only the outline of "Fort Dearborn," with immediately to the south of it a stippled area probably representing fields. The course of the North and South Branches of the Chicago River is easily identifiable; note that on section 30, referring to section 29, are the words "head of navigation." This refers to the point at which the Chicago Portage took off westwards, eventually to reach the Des Plaines River, and thence the Mississippi River. Along this portage would soon be built the canal (the Illinois and Michigan Canal) that would in effect fix this site as the capital of the emergent Midwest.

The only other cultural feature is the dotted and colored line that, for instance, runs alongside the North Branch in section 5 at the top of the plat; this line seems to mark trails, probably those used by the local Indians and trappers. There are very few indications of vegetation type, except that here and there alongside the river, we see the word "timber." Most of the rest of the land would have been either in prairie, or in what the surveyors called *wet prairie*—that is, slough.

It is instructive to compare this early plat with a modern street map of the city of Chicago. The line of the river is our guide in seeing how the various sections were laid out, and it is plain that those sighting lines, established 150 years ago by some deputy surveyors, slogging their way through a most unpromising swamp, have in fact become the main streets of the modern city, their fractional township bounded to the west by Western Avenue, to the north by North Avenue, to the east by the lake, and to the south by Pershing Avenue (was there never a South Avenue?). The accompanying sketch also makes clear how much land has been reclaimed from the lake, much of it after the great fire of 1871.

Source: Robert Knight and Lucius H. Zeuch, The Location of the Chicago Portage Route of the Seventeenth Century *(Chicago: 1928).*

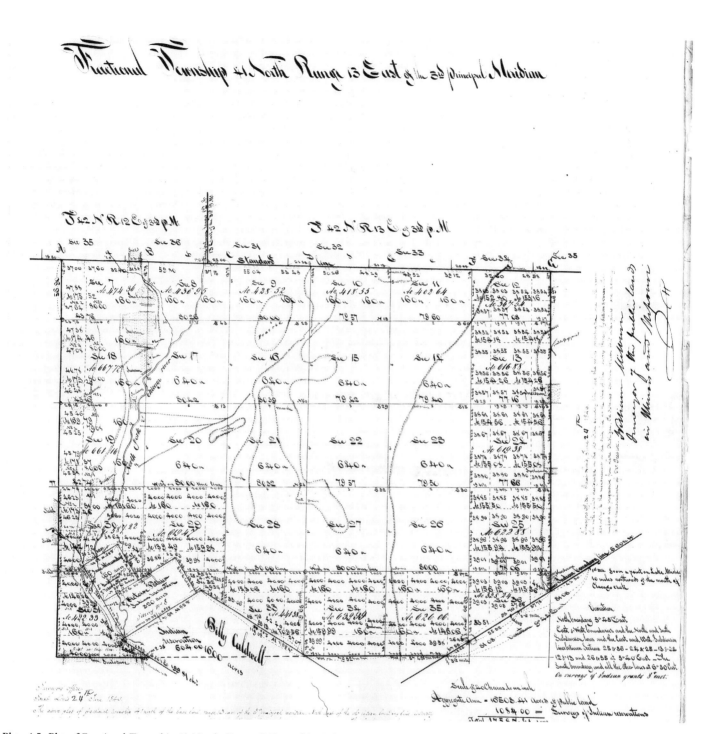

Plate 4.5. Plat of Fractional Township 41 North, Range 13 East of the 3rd Principal Meridian, c. 1836. Illinois State Archives.

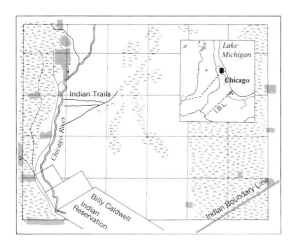

Plate 4.5a. Diagram and
location map for Plate 4.5.

Some Indian Survivals

During 1836 and 1837, the surveyors were busy setting out the townships to the north of Chicago, in the area now occupied by Skokie. They produced for this area a plat that is exceptional for the wealth of its information on both topographical and cultural features. Niles Township, as it was called, was fractional, beginning only at section 7 (top left). The "north fork of the Chicago River" ran down its western side, with some "inclosures" and "houses" marked in stipple alongside it; these were probably the work of early settlers rather than of Indians. Three dotted lines run out from the river eastwards; these probably indicate surviving Indian trails, coming from the river towards the open ground, and thence leading away to the northeast.

Three large plots are marked on the southwestern edge of the plat, and these were the lands alongside the Chicago River assigned to the Potawatomi chief Billy Caldwell, alias Sauganash, by the treaty of 1829. He was the son of an officer in the British service and of a Potawatomi woman, and migrated west with his tribe in 1835. Along the bottom right-hand side of the plat runs the diagonal Indian Boundary Line, established by the federal government in 1816 as the northern edge of a corridor designed to safeguard communication between Lake Michigan and the Illinois River (see the inset map (Plate 4.5a) for an explanation of this).

The plat has a good deal of topographic information. The wandering line running north-south along the right-hand side marks the limit between swampland to the east, and the timber that covered most of the township. Running down from the north are two fingers, with some associated circular features; these are sandy patches where there was prairie rather than woodland. If we combine this plat with the field notes for the area, we learn that the forest was mostly oak-hickory away from the river, but that alongside the Chicago River were maple and basswood.

As is often the case, these features from the early nineteenth century have become fixed in the modern map of this area. The diagonal reservation of Billy Caldwell remains as a noteworthy intrusion into the checkerboard pattern of the streets, and the northeast-trending Indian trail has become a diagonal road. The Indian Boundary Line has here been effaced from the street pattern, but in other parts of Chicago it survives, another incongruous diagonal intrusion upon the checkerboard.

Sources: Gerald Danzer and David Buisseret, Skokie, a Community History Using Old Maps *(Skokie: 1985); and Philip Hanson, "The Presettlement Vegetation of the Plain of the Glacial Lake Chicago in Cook County, Illinois," Ohio Biological Survey Biological Notes XV, pp. 159–64.*

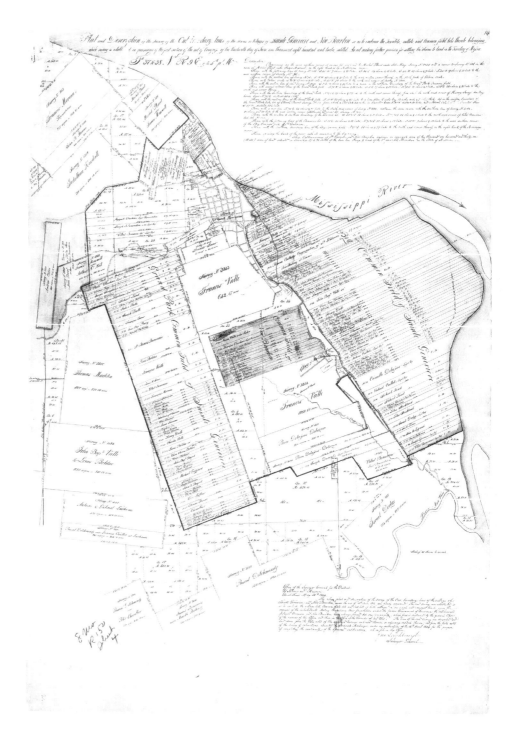

Plate 4.6. "Plat and description of the survey of the outboundary lines of the towns or villages of Sainte Genevieve and New Bourbon," 1854. The National Archives.

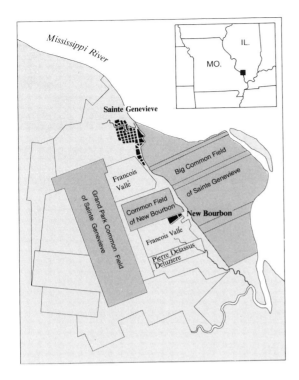

Plate 4.6a. Diagram
of Plate 4.6.

French Intrusions into the Rectangular System

While the plats for most areas record a phase in their settlement under the aegis of the federal government, some depict earlier patterns of land tenure, established under Spanish, French, or British sovereignty. This plat of Townships 37 and 38 North, Range 9 East, Fifth Principal Meridian, around Sainte Genevieve, Missouri, is typical of such maps. As the federal government had early established the principle of honoring valid titles to lands granted by foreign governments, it was important to integrate these into the general township and range system, and the plats that were drawn to achieve this are an excellent source for studying the earlier cadastral patterns that are found in many states outside the township and range system (see Plate 4.1).

Sainte Genevieve, founded about 1740 by the French from Illinois, was the earliest permanent European settlement in Missouri, and this plat plainly shows the French cadastral pattern. The three common fields are divided into long narrow lots, most of which bear the names of French owners, though there are a few anglophone names, reflecting the influx of Americans a few years before the United States acquired this area by the Louisiana Purchase (1804). This long-lot pattern, reminiscent of the medieval European *strips*, was characteristic of French tenurial patterns in the New World, and was probably designed to offer as many people as possible both a share of good land and a frontage onto some waterway.

From this plat we can trace the names of some of the more prominent landholders: François Vallé, civil and military commandant of the village; Jean-Baptiste Vallé, his brother, who replaced him as commandant under the American regime; Israel Dodge, one of the first Americans in the area, who was involved in lead and salt mining nearby; James Maxwell, an Irish priest assigned to the Sainte Genevieve parish in 1796; and Pierre Delassus Deluziere and Camille Delassus, members of the French nobility. The two latter noble names are found near the "village of New Bourbon," which was established in the mid-1790s as a refuge for French nobles who had fled the Revolution in France. New Bourbon did not prosper, however, and the focus of French settlement remained the town of Sainte Genevieve, on its low bluff above the Mississippi River.

Sources: Carl J. Ekberg, Colonial Ste. Genevieve: An Adventure on the Mississippi Frontier *(Gerald: Mo.: 1985); and Pual W. Gates, "Private Land Claims," In* History of Public Land Law Development *(Washington, D.C.: 1968).*

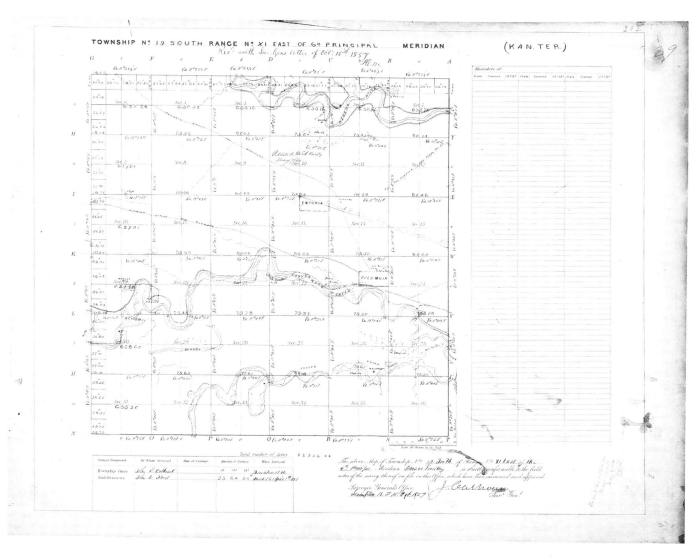

Plate 4.7. Plat of Township 13 South, Range 11 East of the 6th Principal Meridian, 1857. The National Archives.

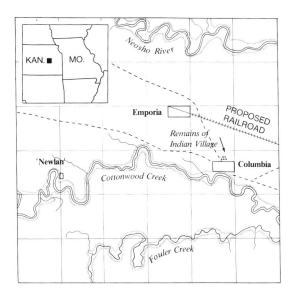

Plate 4.7a. Diagram and location map for Plate 4.7.

Early Settlement in Kansas

By the time the General Land Office surveys reached Kansas, almost seventy years after the initial surveys in Ohio's Seven Ranges, the surveying system was well established. This plat, showing an area roughly midway between Wichita and Kansas City, is typical of the cartographic style that prevailed by the middle of the nineteenth century. It was prepared on a printed form at a scale of one inch to forty chains, and helps document the rapid settlement of eastern Kansas following its establishment as a territory in 1854.

When the township's section lines were surveyed in 1857, there were already several settlers living along the banks of the Neosho River and Cottonwood Creek, as well as on the sites of two proposed towns, Emporia (today a town of 25,000) and Columbia (today extinct). In addition, there was a network of roads converging on Columbia, located on the banks of the Cottonwood, near the site of an abandoned Indian village. Eventually, the railroad (penciled on the plat) came to Emporia, and the irregular road pattern was replaced by roads that followed the township and section lines on their north-south and east-west courses.

The plat is mostly mute on the presettlement pattern of vegetation, but from the accompanying field notes we can learn both about the distribution of the timber and about the stage reached by the various settlements. For example, "S. Griffith's" field is fenced and ploughed and his house is timber; a large swamp extends through most of the two adjoining sections, and the timber along the Cottonwood Creek consists mainly of cottonwood in association with hackberry, elm, oak, ash, maple, and sycamore. Almost all the settlements that are indicated by a square (for buildings) and/or a field pattern (for cultivated areas) are located near the prairie-woodland border along the streams, as we should expect.

Using the tract books and field notes, we can check on the various names, and note that some are misspelled: "John Brown" for Solomon G. Brown, "Newlan" for Newlin, "Fouler Creek" for Fowlers Creek and so forth. However, these misspellings are never serious enough to cast doubt on the identity of the person intended, and the plat as a whole wonderfully catches this township at the very beginning of its development by the settlers.

Source: An Illustrated Historical Atlas of Lyon County, Kansas *(Philadelphia: 1878).*

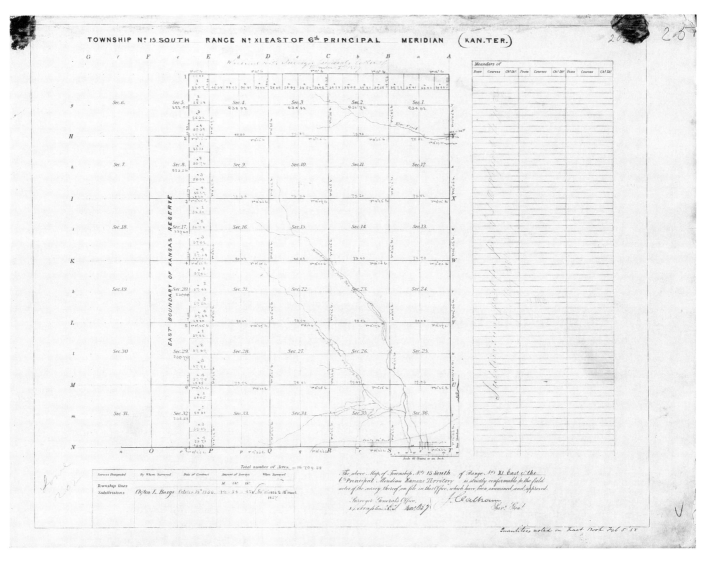

Plate 4.8. Plat of Township 15 South, Range 11 East of the 6th Principal Meridian, 1857. The National Archives.

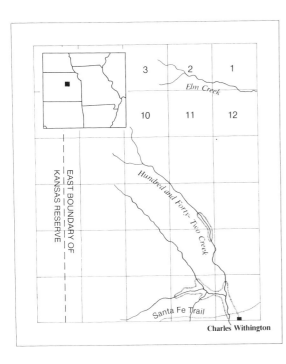

Plate 4.8a. Diagram and location map for Plate 4.8.

Conflicting Claims on the Land

This plat, as we learn from the field notes, covered an area that was predominantly prairie, with the wooded area confined to the banks of Hundred and Forty-Two Creek in the southeastern corner of the township. In 1856 the whole township had only one settler, Charles Withington, whose land straddled the southern township line. His improvements were located along the Santa Fe Trail at its junction with a road heading south to the Neosho Valley. According to an early county historian, Withington was the earliest settler in the area, operating a store—that was also the post office—and helping prospective settlers to find lands, particularly along the Neosho River and Cottonwood Creek.

The western two tiers of sections were not surveyed because this land was part of the reservation set aside for the Kaw or Kansas In-dians; it was to be sold in trust for them by the federal government, and was eventually surveyed in 1861. In the 1870s the Kansas Indians ceded their remaining lands and moved to the Indian Territory, as did other Indian tribes such as the Kickapoo, Potawatomi, Delaware, Sauk and Fox, Ottawa, Osage, and Cherokee; by then the pressure from white settlers was becoming intense.

This area of Kansas was a focus not only for tensions between whites and Indians, but also between pro- and antislavery factions. In fact, this plat was approved by John Calhoun, a strong proslavery politician during the territorial period from 1854 to 1861; ironically, he had taught Abraham Lincoln surveying when he was in Illinois. The Neosho Valley settlements were vulnerable to raids from antislavery men, and in 1856 Withington's store was destroyed in one of these raids. This plat is then a good example of a document that, while at first sight rather dull, in fact proves upon analysis to bear on several of the pressing issues of its day.

Sources: Charles Royce, Indian Land Cessions in the United States, *Bureau of American Ethnology, Eighteenth Annual Report, 1896–1897 (Washington, D.C.: 1899); and Jacob Stotler, "History of Lyon County, Kansas," in* An Illustrated Historical Atlas of Lyon County, Kansas *(Philadelphia: 1878).*

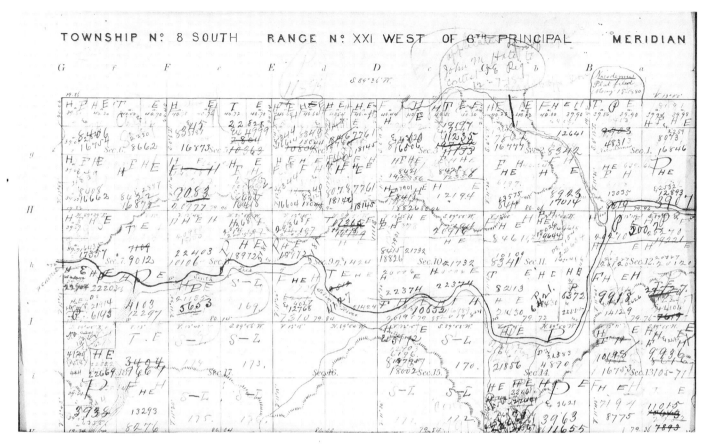

Plate 4.9. Plat of Township 8 South. Range 22 West of the 6th Principal Meridian, 1869. The National Archives.

The Establishment of Nicodemus

This is an example of the triplicate, or local office plat. It is heavily annotated with abbreviations and land entry numbers referring to the transactions involved in the disposition of the land. These annotations obscure the topographic features recorded on the base plat, but in this case the landscape is relatively uniform, consisting of prairie with small wooded areas along the South Fork of the Solomon River.

Although the township was subdivided into sections in 1869, most of the land transactions shown here were not recorded until the late 1870s and early 1880s, when most of the tracts were disposed of through the homesteading process. Section 1 contains the community of Nicodemus, an all-black town settled by freed slaves in the 1870s. Among the associated land entry papers, there is a file on this site, describing the original town plan, that had forty-nine blocks with twenty-four lots each.

Letters of 1884 and 1885 indicate that the town was not growing as rapidly as had

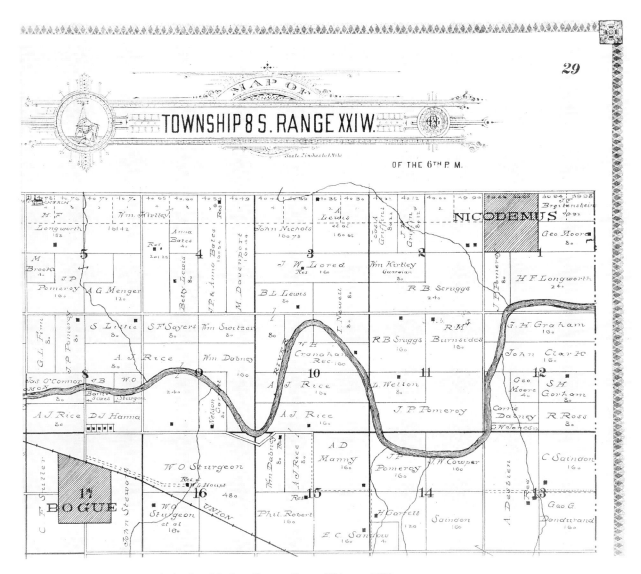

Plate 4.9a. Plat from the *Standard Atlas of Graham County, Kansas* (Chicago, 1906).

been hoped. At that time it had a population of thirty-six inhabitants, fifteen stone, fourteen frame, and seven sod buildings. These buildings included grocery, millinery, furniture, and general merchandise stores, a stone church, two hotels, a post office, bank, stables, barber shop, society hall, and restaurant. Nicodemus never attained the prosperity that its founders hoped for, but it did at least survive, as Plate 4.9a shows, and by the middle of the twentieth century counted about 300 inhabitants.

Source: Paul W. Gates, "Homesteading, 1862–1882," in
History of Public Land Law Development
(Washington, D.C.: 1968).

CHAPTER FIVE

Nineteenth-Century Landscape Views

David Buisseret

In the previous chapter we were dealing with artifacts that anybody would recognize as "maps"; vertical, strictly planimetric representations of the countryside. In this chapter, on the other hand, we deal with material that to some people might not even qualify as a map. But if we define a map as a graphic representation of a locality, then plainly much of what passes for "art" comes into our purview. Rubens' marvelous *Landscape with Castle Steen* is a kind of map, and so is Constable's *Dedham Vale,* to name only two very well-known examples; there is a large area in which the products of art and those of cartography overlap.

Probably the best way of distinguishing between these pictorial genres is by thinking back to the intention of their composers. A map is successful if it conveys to the mind of the reader the salient features of some geographical area, sometimes including its natural or social phenomena. A painting, on the other hand, is only incidentally concerned with conveying topographic information; here, success or failure is gauged by the work's effect upon the aesthetic sensibility of the "reader." To take two extreme examples, a computer-generated map, accurate and useful in its own terms, is

very unlikely to contain any meritorious qualities from an aesthetic point of view, and, at the other extreme, a landscape by Monet, primarily concerned with conveying effects of light, will probably contain little useful information about the actual locality. We are, then, dealing with a wide range of artifacts, some of them shading into what are manifestly not "maps." In this frontier area fall, for instance, paintings with invented landscape backgrounds, which are sometimes described as *imaginary maps*.

The landscape view has a long history in European North America. Topographical views showing the countryside had been produced during the eighteenth century and indeed earlier, for in the first wave of European invasion artists like John White and Jacques Le Moyne (and the anonymous painter of the newly discovered Drake Manuscript at the Pierpont Morgan Library) had captured watercolor images of their New World. Much of the printed material is listed in Lynn Glaser's *Engraved America: Iconography of America through 1800* (Philadelphia, 1970), and this contains a great many views of towns, with some images too of the countryside. However, this early production would be dwarfed by the output of topographical images after about 1830, for reasons that were technical as well as political and cultural.

In political terms, this was the lusty infancy of the new republic, whose sense of national mission, of manifest destiny, cried out for graphic expression. The transformation of the struggling seaboard colonies into states with almost unlimited territorial ambitions meant that the whole region west of the Appalachians was opened up to artists; the conquest of this vast territory offered painters a theme that was extraordinary in its breadth, variety, extent, and romantic appeal. This new political thrust meshed perfectly with a powerful cultural force from Europe, the advent of romanticism in the visual arts, which struck first on the eastern seaboard. Here such regions as the Hudson River Valley and the Maine coast fitted marvelously well into the romantic stereotypes that European painters had formed in the Swiss Alps and along the English coast; the painters of eastern North America thus developed schools that were poised to take advantage of the opening of the West. To these powerful political and cultural trends were added new technical developments. During the last decades of the eighteenth century, advances in paint manufacture and in papermaking in England encouraged British artists to develop the technique of painting with transparent colors on specially made white paper, and these techniques rapidly crossed the Atlantic, so that we find many of the leading artists of North America at ease both with the new watercolors and with the long-established technique of oil painting.

Whichever medium the artists chose, two other technical developments made it possible to disseminate their paintings more widely than ever before: lithography and steel engraving. Lithography, invented in Bavaria at the end of the eighteenth century, became common in the United States during the 1830s. It offered the possibility of making duplicate images of great subtlety at very competitive prices, and soon had largely supplanted older techniques such as copper engraving. The latter process was relatively expensive, as the plates could take only a limited number of impressions (or *strikes*). Toward the middle of the nineteenth century techniques of engraving on steel were developed, and these steel plates greatly reduced production costs, as they could sustain very long runs without losing their sharpness of detail. In a sense the two new techniques were complementary, as lithography could render delicate tonal effects, while steel engraving was unrivaled for conveying sharp detail.

One way and another, then, the first half of the nineteenth century was very fertile in its introduction of new processes of image-making. These processes came to serve immense new markets, which emerged not only because of political and cultural shifts, but also because of the massive growth of the population and of the economy. By the middle of the century, a substantial proportion of the U.S. population was able and willing to acquire books containing descriptions and topographical views of their country. Indeed, the output of these views eventually became so great that it is hard to impose any sort of pattern on so vast a production.

Faced with this great abundance, I have tried to choose images from a wide variety of techniques and geographical areas, selecting those of most interest to historians, and those of which good reproductions are readily available. As with any other type of evidence, we have to be on the lookout for problems of accuracy, and with some images we encounter a novel type of persuasion, when the aesthetic message can be almost overwhelming. Topographical views are of course like any other type of visual evidence, in that they require close and patient study, for which either originals or good reproductions are essential.

For almost the first third of the nineteenth century, we have little to record. But then around 1830 the flood begins. Many of the early topographical artists came from England, like John William Hill (1812–1897), whose delicate watercolors and aquatints offer charming images of the early industrial East (see Plate 5.1). However, they were soon joined by native-born artists such as George Catlin (1796–1872). Catlin is best known for his images of Indian life, for which he enlisted on innumerable expeditions from 1830 onward, but he was also an excellent landscape artist who caught the West at the moment when it was just being penetrated by Europeans (Plate 5.2). He eventually accumulated a huge body of graphic information about Indian life, known as the Catlin Collection. Immediately after Catlin's first images come those of Karl Bodmer (1809–1893), a Swiss who had trained in Europe and between 1832 and 1834 accompanied Prince Maximilian of Wied on a great journey in the West. Bodmer never returned to the New World after this journey, but his highly developed romantic sensibility allowed him to capture during his travels images that are astonishingly evocative, both in the watercolor originals and in lithographic reproductions (Plate 5.3).

While Catlin and Bodmer were working in the foothills of the Rocky Mountains, Seth Eastman (1808–1875) was taking up his first posting as a soldier, at Fort Crawford, near Prairie du Chien, Wisconsin. Moving from there to Fort Snelling, near Minneapolis, he began a life of great artistic productivity, much of it in collaboration with his wife, Mary. He was for a while assistant teacher of drawing at the U.S. Military Academy at West Point, and contributed the illustrations for Henry Schoolcraft's *Historical and Statistical Information Respecting the Indian Tribes of the United States* (6 vols., Philadelphia, 1851–1857). Eastman's work was characterized by great accuracy and by strong feeling for the Indians who were so often his subjects and his friends.

By the late 1830s it was no longer unusual for artists to go West in search of subjects.

Alfred Joseph Miller (1810–1874) in 1837 accompanied Captain William Drummond Stewart on a trading voyage from Saint Louis to the Oregon Territory and back, producing firsthand but rather poetic images. In 1842 John Mix Stanley (1814–1872) began his very prolific career at St. Louis. He painted widely in the South and Southwest, and then went to Oregon and Hawaii. In 1853 he was the artist for the railroad expedition that investigated the route west on the 47th and 48th parallels. As a collector, he was in a sense the successor of Catlin, building up a great mass of material that was largely destroyed in a fire at the Smithsonian Institution. His work is remarkable for its accuracy as well as its elegance.

One of the earliest artists in the Southwest was Theodore Gentilz (1820–1906), who trained in Paris and was then engaged in 1843 by Henri Castro to depict the latter's colony in Texas. Gentilz produced evocative images of Comanche life, and views of mission stations and of towns like Brownsville. While Gentilz was working in Texas, the far northern frontier of the United States was being depicted by Henry Warre (1819–1898). Warre was a British officer who between 1844 and 1846 worked his way from Montreal to the mouth of the Columbia River and back, making delicate watercolor and pencil drawings of sites on the route, many of which he published in *Sketches in North America and the Oregon Territory* (London, 1848). A few years later Paul Kane (1810–1871) covered much of the same ground. After studies in Europe—where he probably met and was inspired by Catlin—Kane returned to Toronto and set to work depicting central and western Canada, publishing his work in *Wanderings of an Artist* (London, 1859).

During the 1840s and 1850s, the great westward migration, largely along the Oregon Trail, inspired many artists. Some of the most striking images were produced by James F. Wilkins (1808–1888), who had come to the United States from England in 1837, and in

1849 left St. Louis bound westwards. His delicate pen-and-ink sketches leave an extraordinary impression of wagons fording rivers, wagons crossing vast plains, and wagons descending quite improbable gullies, the very understated nature of his art carrying all the more conviction. While Wilkins was sketching on the Oregon Trail, Captain Arthur T. Lee, Eighth U.S. Infantry, was busy with his carefully crafted watercolors of West Texas, and later of the upper Mississippi. Lee was a worthy successor to Eastman and Warre, in an age when soldiers learned to sketch as part of their basic training.

Leaving the West and its artists, we come back to the settled East, where George Inness (1825–1894) was one of the artists of the late Hudson River School. He was commissioned by the president of the Delaware, Lackawanna, and Western Railroad to paint scenes along its path, and in 1855 produced the justly famous *Lackawanna Valley,* at a time when he had not escaped the realism of the Hudson River School but was beginning to paint with the abstract symbolism of his later years (Plate 5.7). Inness worked within what we might call the grand tradition, closely influenced by contemporary artistic developments in Europe. So did Sanford Robinson Gifford (1823–1880), one of whose oil paintings we have chosen because of its powerful evocative force (Plate 5.12). Gifford was born in New York state, and after studying in Europe returned to work primarily in the East, becoming particularly well known for his studies of the Catskills and Berkshires. A younger contemporary was William Henry Jackson (1843–1942). Jackson had a stupendously productive life, first in sketches and paintings, and then in photographs. Much of his work has been reproduced in books by his son, Clarence Jackson (see the bibliography at the end of this chapter). We have here chosen to show his *California Crossing, South Platte River* (Plate 5.11) because of its brilliant evocation of the scenes along the Oregon Trail.

The nineteenth century was a great age for artists who wished to publish their work in lithographs or steel engravings. One of the earliest such publications was *American Scenery,* published by N. P. Willis in London in 1840. The two volumes contain 119 plates, from drawings by the Englishman W. H. Bartlett. The subjects include buildings and town views as well as landscapes, the latter particularly interesting as images of the early industrialization of North America (Plates 5.4 and 5.5). When Charles Dana published *The United States Illustrated* (New York: 1855), the machines and even the countryside had already changed. Dana's thirty-three views are engraved with a particularly meticulous touch, so that his details are exceptionally clear (Plate 5.6). The engravings in Robert Sears' *A Pictorial Description of the United States* (New York: 1855) are not so sharply defined, but they are extremely numerous, and cover every part of the Union (Plate 5.8).

During the 1840s, the federal government began to sponsor expeditions to the remoter parts of North America, in order to define boundaries, enumerate natural resources, identify potential railroad routes, and so forth. Many of the reports from these expeditions (listed in Hasse in the bibliography at the end of this chapter) contain remarkable topographical views. This is particularly true of the railroad surveys, from which examples are reproduced as Plates 5.9 and 5.10. Alongside this governmental activity we still find commercial publications such as *Picturesque America,* edited by William Bryant Cullen (2 vols., New York, 1872 and 1874), and the most successful of them all, the great series of prints published by Currier and Ives. After the middle of the century large numbers of topographical views appeared in periodicals such as *Harper's Weekly,* though these were eventually superseded by photographs. The coming of the photographic age produced, indeed, so huge a mass of material that I have recoiled from attempting to give any account of it. Still, if we count only the material produced from about 1830 to the advent of photography, we have a large and steadily increasing volume of views, many of which are of great interest to the historian.

SOURCES FOR STUDYING TOPOGRAPHICAL VIEWS

GENERAL HISTORIES AND CATALOGS OF TOPOGRAPHICAL VIEWS

Abbey, J. R. *Travel in Aquatint and Lithography from the Library of J. R. Abbey.* 2 vols. London: 1956 and 1957. Full listings of the main works in Abbey's library from all over the world; the plates are excellent but few.

Bakeless, John. *The Eyes of Discovery: America as Seen by the First Explorers.* New York: 1961. This is a fascinating text, but the book is marred by having illustrations that are too few and too small.

Born, Wolfgang. *American Landscape Painting: An Interpretation.* New Haven: 1948. This survey examines the European roots of the genre, and offers a convincing typology of the different styles and schools.

Curry, Larry. *The American West: Painters from Catlin to Russell.* New York: 1972. An exhibition catalog containing good plates from the work of the major Western painters.

Czestochowski, Joseph S. *The American Landscape Tradition: A Study and Gallery of Paintings.* New York: 1982. This short survey has a wealth of color plates, each one with a helpful commentary.

Davidson, Marshall B. *Life in America.* 2 vols. Boston: 1951. A tremendous compendium of visual images, many of them not well known, reproduced in small black-and-white photographs.

Deák, Gloria Gilda. *Picturing America, 1497–1899.* 2 vols. Princeton, N.J.: 1988. Prints, maps and drawings bearing on the New World discoveries and on the development of the territory that is now the United States.

DeVoto, Bernard. *Across the Wide Missouri.* Boston: 1947. Early but well-illustrated account of the painting of the West.

Eliot, Alexander. *Three Hundred Years of American Painting.* New York: 1957. Interesting chapters on the Hudson River School, George Inness, and the Western painters.

Ewers, John C. *Artists of the Old West.* New York: 1965. Systematic survey of painters from the time of Lewis and Clark, with each chapter devoted to one or two painters.

Gardner, Albert Ten Eyck. *A History of Watercolor Painting in America.* New York: 1966. Particularly interesting on the technological background to cultural developments.

Glanz, Dawn. *How the West Was Drawn: American Art and the Settling of the Frontier.* Ann Arbor: 1982. An interesting attempt to relate artistic developments to such themes as the conflict between civilization and wilderness, or the U.S. sense of national mission.

Glaser, Lynn. *Engraved America: Iconography of America through 1800.* Philadelphia: 1970. Good background material for understanding the greatly increased production of the nineteenth century; it has many topographical views from the seventeenth and eighteenth centuries.

Goetzmann, William H. and William N. Goetzmann. *The West of the Imagination.* New York: 1986. An interesting survey of Western painters; the plates are numerous and in color, but rather small.

Hasse, Adelaide Rosalia. *Reports of Explorations Printed in the Documents of the U.S. Government.* Washington, D.C. 1899. Still a very useful finding aid for graphic material.

Hassrick, Peter. *The Way West: Art of Frontier America.* New York: 1977. This book resembles the Goetzmann volume, except that here the text is shorter and the plates better reproduced.

Keaveney, Sydney Starr. *American Painting: A Guide to Information Sources.* Detroit: 1974. A very useful manual, with headings on individual artists and on museums and important collections.

McCracken, Harold. *Portrait of the Old West.* New York: 1952. Another survey of Western art, with chapters on individual artists.

Moritz, Albert F. *America the Picturesque in Nineteenth-Century Engraving*. New York: 1983. A good survey of this neglected area, putting the artists into the context of the engravers.

Novak, Barbara. *American Painting of the Nineteenth Century*. New York: 1969. This is an interesting attempt to define the "Americanness of American art."

Peters, Henry T. *America on Stone*. New York: 1931. An alphabetical list of lithographers with 153 plates, only some of which show landscapes.

Ruth, Kent. *Great Day in the West: Forts, Posts, and Rendezvous beyond the Mississippi*. Norman, Okla. 1963. A brilliant guide to Western places, juxtaposing old prints and modern photographs.

Stebbins, Theodore. *American Master Drawings and Watercolors*. New York: 1976. A masterly survey, but a little disappointing from our point of view, as there are relatively few landscapes in these mediums.

TOPOGRAPHICAL VIEWS BY SUBJECT OR AREA

Benes, Peter. *New England Prospect: A Loan Exhibition of Maps at The Currier Gallery of Art*. Boston: 1981. Catalog of an exhibition making very clever use of early maps and views in its analysis of New England life.

Howat, John K. *The Hudson River and Its Painters*. New York: 1972. A fine example of a detailed study, with about 100 well-commentated plates, as well as modern maps and photographs.

Jakle, John A. *Images of the Ohio Valley: A Historical Geography of Travel, 1740 to 1860*. New York: 1977. The imagery is mostly verbal, but the few prints are accompanied by very interesting commentary.

Peters, Henry T. *California on Stone*. New York: 1935. A list of lithographers, with reproductions of some of their works, which include landscapes.

Rathbone, Perry T., ed. *Mississippi Panorama*. St. Louis: 1950. A tremendous collection of images from the length of the Mississippi, not very well reproduced.

Rathbone, Perry T., ed. *Westward the Way*. St. Louis: 1954. A companion volume to *Mississippi Panorama* and, like it, a great source for landscape images.

Van Nostrand, Jeanne. *The First Hundred Years of Painting in California*. San Francisco: 1980. Splendid selection of regional views, with a commentary on their artists.

Van Zandt, Roland. *Chronicles of the Hudson: Three Centuries of Travellers' Accounts*. Rutgers: 1971. The author added many topographical views alongside the accounts. It is a fine example of the way in which these types of evidence complement each other.

Walther, Susan Danly, ed. *The Railroad in the American Landscape, 1850–1950*. Wellesley, Mass.: 1981. Excellent analyses of landscapes, supplemented by relevant photographs.

TOPOGRAPHICAL VIEWS BY COLLECTION

American Paintings from the Metropolitan Museum of Art; Catalog of an Exhibition held at Los Angeles County Museum of Art in 1966. Los Angeles: 1966. Offers an overview of this fine collection.

Amon Carter Museum of Western Art: Catalogue of the Collection 1972. Fort Worth, 1973. Six hundred pages of material, with small photographs of some very interesting topographical views.

Deák, Gloria-Gilda. *American Views: Prospects and Vistas*. New York: 1976. A collection of 128 views from the New York Public Library collections, with brief commentary.

Ewers, John C. *Views of a Vanishing Frontier*. Omaha: 1984. Catalog of the InterNorth Art Foundation's magnificent holdings of Bodmer material.

Fairbanks, Jonathan. *Frontier America: the Far West*. Boston: 1975. Catalog of an exhibition mounted by the Museum of Fine Arts, Boston, containing some fine and little-known topographical material.

Goetzmann, William H., and Joseph C. Porter. *The West as Romantic Horizon*. Omaha: 1981. Selections from the collection of the InterNorth Art Foundation.

Hassrick, Peter. *Treasures of the Old West: Paintings and Sculpture from the Thomas Gilcrease Institute of American History and Art*. New York: 1984. This catalog of a splendid collection is exceptionally well produced, with fine plates.

Hunt, David C. *Legacy of the West*. Omaha: 1982. Offers a conspectus of the Joslyn Art Museum's Western collections, including many topographical views.

Rossi, Paul A., and David C. Hunt. *The Art of the Old West*. Tulsa: 1971. Well-illustrated selections from the Thomas Gilcrease Institute of American History and Art in Tulsa, Oklahoma.

LIVES AND WORKS
OF INDIVIDUAL ARTISTS

Bell, Michael, ed. *Brave and Buffalo: Plains Indian Life in 1837*. Toronto: 1973. Watercolors of Alfred Jacob Miller.

Cikovsky, Nicolai. *George Inness*. New York: 1971.

Goetzmann, William H., ed. *Karl Bodmer's America*. Omaha: 1984. These watercolors are reproduced with remarkable fidelity.

Hanna, Archibald, ed. *Sketches in North America and the Oregon Territory by Captain H. Warre*. Barre, Mass.: 1970.

Harper, J. Russell. *Paul Kane's Frontier*. Austin: 1971.

Hendricks, Gordon. *Albert Bierstadt, Painter of the American West*. New York; c. 1975.

Jackson, Clarence S. *Picture Maker of the Old West: William H. Jackson*. New York: 1947.

Kendall, Dorothy Steinbomer. *Gentilz: Artist of the Old Southwest*. Austin: 1974.

McCracken, Harold. *George Catlin and the Old Frontier*. New York: 1959.

McDermott, John Francis, ed. *An Artist on the Overland Trail: The 1849 Diary and Sketches of James F. Wilkins*. San Marino, Calif.: 1968.

————. *Seth Eastman: Pictorial Historian of the Indian*. Norman: 1961 and McDermott, *Seth Eastman's Mississippi: a Lost Portfolio Recovered*. Urbana, Ill.: 1973.

Ross, Marvin C. *The West of Alfred Jacob Miller*. Norman, Okla.: 1968.

Thomas, Davis and Karin Ronnefeldt, eds. *People of the First Man*. New York: 1976. Bodmer watercolors.

Thomas, W. Stephen. *Fort Davis and the Texas Frontier: Paintings by Captain Arthur T. Lee, Eighth U.S. Infantry*. College Station, Texas: 1976.

Truettner, William H. *A Study of Catlin's Indian Gallery*. Washington, D.C.: 1979.

Plate 5.1. *View on the Erie Canal;* watercolor drawing by J. W. Hill, 1830–32. I. N. Phelps Stokes Collection, Miriam and Ira D. Wallach Division of Art, Prints and Photographs, The New York Public Library, Astor, Lenox and Tilden Foundations.

An Early Image of a Crucial Communications Link

The Erie Canal, connecting New York to the Great Lakes, had been opened in 1825, and is here shown in a typical watercolor. A team of three mules tows a good-sized barge on the canal, and a smaller barge is apparently hitched to it. In the smaller barge we can see piles of stone. The larger one looks like a passenger barge or *packet,* with its numerous occupants and high superstructure.

The countryside has a well-tended, settled look, as would be appropriate for this part of New York state around 1830. Note, though, that the artist does not hesitate to draw exactly what he sees, even if this gives an impression of

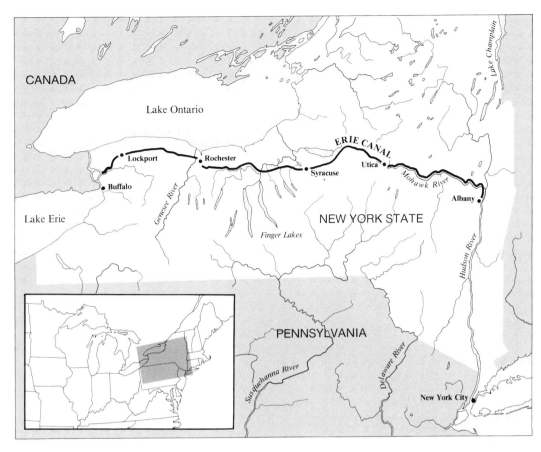

Plate 5.1a. Modern map showing the course of the Erie Canal, with location map.

some degree of dilapidation; some of the fences are decidedly broken down, and the road that crosses over the canal is evidently very uneven. As we shall see, these details tend to get lost in the process of making lithographs and engravings, whose eventual impression is more orderly and so less realistic than the images caught on the wing by the watercolor painter.

By linking Albany with Buffalo, the Erie Canal made possible a great spurt of economic development. It also introduced a quite new element into maritime traffic. Henceforward, views of New York harbor would show not only the graceful oceanic sailing vessels, but also these stubby boats from the interior, all forming part of a vast communications system that eventually would open up the heartland. In fact, this pastoral scene is deceptive, like most such canal scenes. Only when artists showed lock systems, as in Plate 5.5, do we get any idea of the immense resources that went into this kind of construction, for once water fills the canal, the scope of the work is hidden. Canals in fact took enormous effort, both politically and economically, which is why all the largest ones had to be the result of state and federal cooperation.

Source: George E. Condon, Stars in the Water: The Story of the Erie Canal *(New York: 1974).*

Plate 5.2. *Saint Louis from the River Below;* oil on canvas by George Catlin, 1832–33. National Museum of American Art, Smithsonian Institution. Gift of Mrs. Joseph Harrison, Jr.

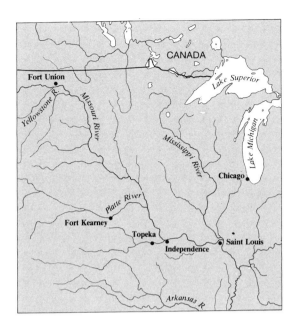

Plate 5.2a. Modern map showing Catlin's route to Fort Union.

Early Penetration of the West: Yellow Stone Leaves Saint Louis

In this early view of St. Louis, Catlin shows the side-wheeler *Yellow Stone* just below, that is south, of St. Louis, on the Mississippi River. Smoke is pouring from her funnels, and steam from her whistle, as she goes on a proving run. Eventually, stern-wheelers proved more effective in the kind of snag-ridden, low-water navigation that *Yellow Stone* was about to undertake, but this had not yet been appreciated. In the background is the town, which at this time consisted merely of seven or eight rows of houses alongside the river. There is as yet no sign of the levee, for the line of steamboats is simply drawn up alongside a fairly high cliff.

Yellow Stone left with Catlin late in March 1832, and after many vicissitudes along the Missouri River succeeded in reaching the American Fur Company's fort at the mouth of the Yellowstone River, roughly two thousand miles northwest from St. Louis. Here, as Catlin wrote, the "Fur Company have erected a very beautiful Fort for their protection against the savages." This was Fort Union, erected in 1829, and destined to be often portrayed by Western artists (Plate 5.3).

Sources: James B. Musick, Saint Louis as a Fortified Town *(St. Louis: 1941); and Francis Paul Prucha,* A Guide to the Military Posts of the United States 1789–1895 *(Madison: 1964).*

Plate 5.3. "Fort Union," from
Prince Maximilian
Wied-Neuwied's *Reise in das
innere Nord-Amerika* (2 vols.,
Coblenz, 1839 and 1841).

A European Stronghold in Indian Country

Karl Bodmer and Prince Maximilian reached Fort Union in 1833, after a journey of seventy-five days from St. Louis. Built in 1828, Fort Union was a formidable 200-foot-square structure, with 20-foot wooden walls and two taller stone bastions. As Plate 5.2a shows, it stood at the confluence of the Yellowstone River with the Missouri River, and in the 1840s was the most westerly stronghold of the American Fur Company.

The two stone bastions and the substantial internal buildings are visible on Bodmer's aquatint, which is a faithful reproduction of his original sketch (see *Karl Bodmer's America*, p. 191, referenced at the end of this plate description). It is taken from the north, looking southwards towards the fort and behind it to the river and the low bluffs on the southern side. In the foreground are some of the Assiniboin Indians who traded at the post; they are breaking up their camp and moving off, carrying some of their goods on the travois behind the white horse.

This aquatint is not only a convincing representation of an actual historic moment, but also contains powerful symbolism and a remarkable atmospheric quality. The four-square fort stands for the uncompromising intrusion of the Europeans into the territory; around it are the fragile-looking and dispersed tepees of the Indians, looking as if they could offer little resistance to this powerful intruder, over whose stronghold flies an enormous stars and stripes.

In fact, for about three decades this fort ensured the survival and prosperity of the American Fur Company in the area. By the time that the United States Army arrived there, in 1864, it had almost collapsed, like the fur trade that it had protected. Today its site is marked only by a few cellar pits.

Sources: William H. Goetzmann, ed., Karl Bodmer's America (Joslyn: 1984); and Kent Ruth, Great Day in the West: Forts, Posts and Rendezvous beyond the Missouri (Norman: 1963).

Plate 5.4. "Rail-road Scene, Little Falls," from *American Scenery,* ed. N. P. Willis (2 vols., London, 1840).

Distance Conquered: An Early Railroad Train

This plate from *American Scenery* takes us back to the early days of the railroad in North America. The engine, of a very early type, is pulling a short string of carriages along the valley between Utica and Little Falls, in New York state. The engraver shows a distinctly European feel to his work, particularly in the delineation of the picturesque town above the falls. Often, the commentary in these collections of engravings has an originality and pithiness that is irresistible:

Before the completion of the railroad, when travellers to the West were contented with the philosophic pace of the canal-boat, one might take up a novel at Little Falls, and come fairly to the sequel by the time the steersman cried out "bridge" at Utica. There were fifteen miles between them in those days; but now (to a man of indistinct ideas of geography, at least, and a traveller on the railroad) they are as nearly run together as two drops on a window-pane. The intermediate distance is, by all the usual measurements of wear and time, annihilated . . . All this is very pleasant to people in a hurry; and as most people in our busy country come under that category, it is a very pleasant thing for the white man altogether. There is a class of inhabitants of the long valley of the Mohawk, however, of whose sufferings, by the advance of the white man's enterprise, this is not the first, though it may be the least, and last. The poor half-naked Oneida, who ran by the side of the once crowded canal-boat for charity, has not time, while the rail-car passes, even to hold out his hand!

Here we have a contemporary reflection upon the beginning of that process, almost indispensable to the growth and survival of the Republic, by which travel time was progressively telescoped, the railroad representing the first and most essential element in that process. We also have a most practical example of the way in which the Indians were progressively excluded from life in the newly settled areas, not always by main force, but often by simple marginalization, so that they just ceased to exist in the minds of the new masters, except as a nuisance to be eradicated.

Source: N. P. Willis, ed., American Scenery, 2 vols. (London: 1840).

Plate 5.5. "Lockport, Erie Canal," from *American Scenery,* ed. N. P. Willis (2 vols., London, 1840).

Overcoming the Barriers to the West: Lockport, 1840

Lockport, shown on Plate 5.1a, was the site of a huge series of locks, which allowed boats to overcome the natural barrier of the Niagara Falls. There were two separate flights of five locks each, and in this engraving the artist is looking down from the top of this huge incline. The locks all had to be opened by hand, and it looks as if the hatted figures are perhaps wearing their lockhands' uniforms.

The editor of *American Scenery* was a great enthusiast for canal travel, as he shows in this passage:

The packet-boats are long drawing-rooms, where (the traveller) dines, sleeps, reads, lolls or looks out of the window; and if in want of exercise, he may at any time get a quick walk on the tow-path, and all this without perceptible motion, jar or smell of steam. Of all the modes of travelling in America, the least popular and the most delightful, to our way of thinking, is travelling on the Canal.

Alas, this delightful mode of travel rapidly gave way to the swift convenience of the railroads. But while the canal enthusiasm endured, it gave rise to some remarkable aspirations and forcibly enunciated expressions of faith in the future, like this extract from *Niles' Weekly Register* for 6 August 1814:

By the Illinois River, it is probable that Buffalo, in New York, may be united with New Orleans, by inland navigation, through Lakes Erie, Huron, and Michigan, and down that river into the Mississippi! What a route! How stupendous the idea! How dwindles the importance of the artificial canals of Europe, compared with this water communication.

So as we look at the Lockport locks, we should see not only a remarkable engineering feat, but also the product of a political will that would be content with nothing short of manifest destiny.

Source: George E. Condon, Stars in the Water: The Story of the Erie Canal *(New York: 1974).*

Plate 5.6. "Wheeling in Virginia," from *The United States Illustrated,* ed. Charles Dana (New York, 1855?).

Frontier City to the West, Wheeling 1855

Charles Dana (1819–1897) was an enterprising journalist and newspaper editor who had access to the best engravers of the day. The view of Wheeling published in his *The United States Illustrated* (New York: 1855) is a fine example of their work in its meticulous detail. The view looks down on the town from the north, with the Ohio River stretching away to the south. On the right is part of Ohio, in fact the part that

was first surveyed after the Land Ordinance, in 1786 (see Plates 4.1 and 4.3).

The engraving neatly accompanies the detail from Charles Augustus Mitchell's map of 1852, which shows "Virginia with its canals, roads and distances from place to place, along the stage and steam boat routes." We are left in no doubt about the importance of these two modes of transport, for the river has two side-

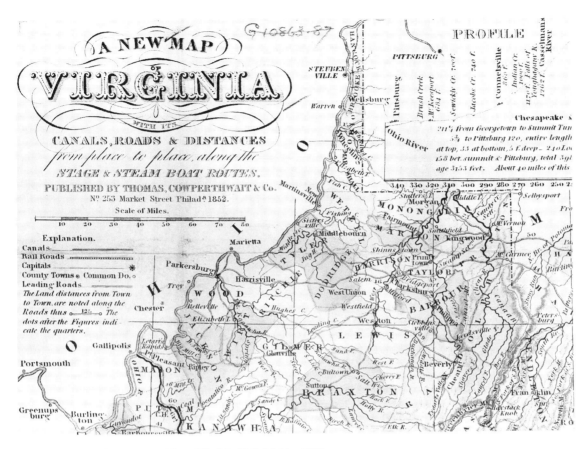

Plate 5.6a. Detail from *A New Map of Virginia* (Philadelphia, 1852).

wheelers (and a flatboat) on it, and a stage coach is just coming up the hill out of the town, to the north. Note also the carts and carriage crossing the Ohio River on the elegant suspension bridge.

There is some hint of industrial activity in the two smoking chimneys away to the south, but on the whole the impression is of an idyllic little country town set in an unspoiled countryside. On his map, Mitchell has provision for marking railroads by the usual conventional symbol, but at this time the nearest line was apparently the one away in the east, leading through Harper's Ferry to Baltimore.

Dana's commentary sets Wheeling into its historical context. At one time, he says, it

looked like becoming a metropolis, but "the increase of Pittsburgh upon the north, and Cincinnati to the southward," has ended these hopes. Moreover, some of Wheeling's importance lay in its position upon the "great national road," but this has by 1855 itself become "an enterprise of the past, which becomes Lilliputian in its proportions of projection, stretched by the side of the Western railroad, which is destined before many years to urge its iron sinews into the surf of the Pacific Ocean."

Source: Charles Dana, The United States Illustrated *(New York: 1855).*

Plate 5.7. *The Lackawanna Valley,* oil painting by George Inness. National Gallery of Art, Washington. Gift of Mrs. Huttleston Rogers.

Industrialization: The First Intimations of Paradise Lost

This painting has long attracted commentary because of its artful blending of nature and artifice. George Inness (1825–1894) was an established painter in the traditional mold when in 1854 he was asked by George Phelps, first president of the Delaware, Lackawanna and Western Railroad, to come and paint scenes along the railroad's route. He painted a brilliant image of the *Delaware Water Gap,* in which the role of

the railroad was very muted, but achieved remarkable fame with this extraordinary view of one of the line's three small locomotives emerging from the yards at Scranton, Pennsylvania.

The map, Plate 5.7a, taken from the *Historical, Topographical Atlas of the State of Pennsylvania* (Philadelphia: 1872), shows the scene roughly fifteen years later. The painting seems to have been made from about the "S" of the south indicator, overlooking the triangular junction, and looking south to the very prominent roundhouse (where rolling stock was turned round on a great turntable, an essential part of early railroads). In the painting, the fields are as yet bare of streets, and it is this

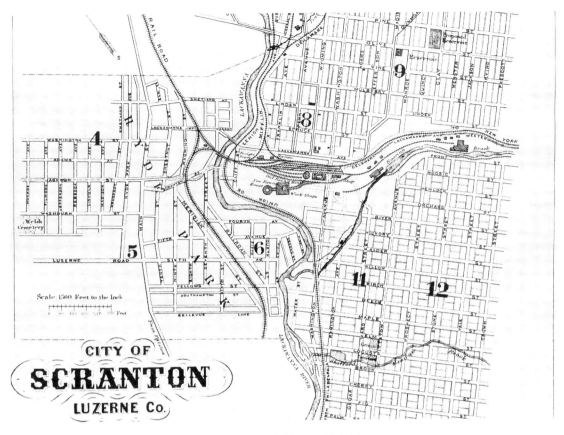

Plate 5.7a. Detail from "City of Scranton" in the *Historical, Topographical Atlas of the State of Pennsylvania* (Philadelphia, 1872).

sylvan air that has so caught the attention of commentators; as Susan Walther puts it, it is a remarkable example of Inness' ability to "reattach even artificial things to nature by a deeper insight." The process of reattachment is achieved through such subtle touches as the gentle curves of the line, which echo those of nature, or the white billows of the engine's smoke, which seem to be an extension of the smoke rising from the town.

The time is probably early morning; there is a hazy quality about the light, which floods in from the east (left). The theme of industrialization, and even despoliation, is far from having gained the upper hand. But the boy in the foreground may stand for primal innocence in the face of advancing civilization, and the tree stumps certainly seem to presage a

destruction of the balance between humans and nature. As always in the presence of a masterly image, we are forced to pay attention not only to the documentary aspect of the scene, but also to its deeper historical meanings.

Sources: Robert J. Casey and W. A. S. Douglas, The Lackawanna Story: the First Hundred Years of the Delaware, Lackawanna and Western Railroad *(New York: 1951); Leo Marx,* The Machine in the Garden *(New York: 1964); and Susan Danly Walther, ed.,* The Railroad in the American Landscape, 1850–1950 *(Wellesley, Mass.: 1981).*

Plate 5.8. "View of Chicago from the Prairie," from *A Pictorial Description of the United States,* ed. Robert Sears (New York, 1855).

Chicago in 1855:
A City in the Prairie

Robert Sears (1810–1892) was a New York publisher who was a skillful advertiser and a great patron of the wood engravers. This view of Chicago, from *A Pictorial Description of the United States* (New York: 1855), is a good example of the types of view that Sears included in his publications. The city is seen from the west, roughly from where it says "Railroad" on the map of 1851 by James H. Rees (though the artist apparently makes no mention of the railroad).

Rees shows the city as platted for eighteen blocks north-south, and twelve east-west,

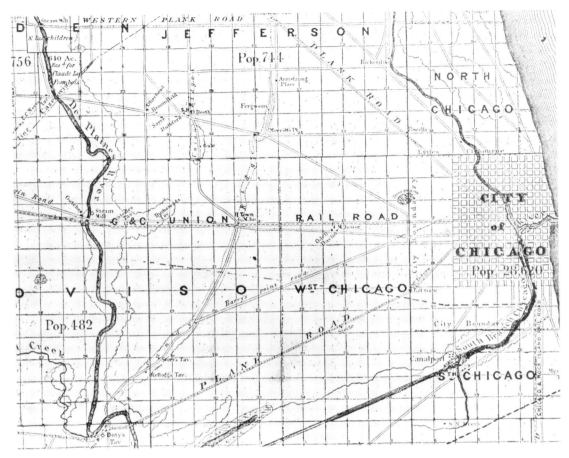

Plate 5.8a. Detail from James H. Rees, *Map of the Counties of Cook and Dupage* (Chicago, 1851).

but according to the plate this area was not yet covered by buildings, since the prairie apparently came right up to the south branch of the Chicago River (on the right in the engraving, and running southward out of the town on the map). The forest of masts in the center of the engraving marks the mouth of the Chicago River, around which the buildings seem as yet to be disposed rather haphazardly.

The engraving seems to show the city as looking rather *too* bucolic for as late as 1855, and Sears may have been using an image that had been made some time earlier. On the other hand, the figures for population of townships on the map are not high. As we shall see in the chapter on city views and plans, the only way to check on the accuracy of such material is to assemble a chronological list of all available images, from which it will be possible to relate the topographical views to the contemporary maps and printed texts. Such an exercise is an invaluable one for students, since it obliges them to deal at first hand with problems of verification and accuracy.

Source: Alfred T. Andreas, History of Chicago, *3 vols. (Chicago: 1884–1886).*

Plate 5.9. "Camp Scene in the Mojave Valley of Rio Colorado," from the *Reports of Explorations and Surveys . . . 1853–1854* (vol. 3, Washington, 1856).

The Republic Strikes West: The Mojave Valley in 1856

During the 1850s various expeditions were mounted to determine the best routes for a railroad to the Pacific Ocean. One of these expeditions went upon "the route near the 35th parallel." Its report was written by Lt. A. W. Whipple, and published in the *Reports of Explorations and Surveys*. Like most of these reports, Whipple's contained not only lithographs of scenes along the route, but also maps identifying each camp site. Whipple's party left from Fort Smith, on the Arkansas River, on July 14,

1853. They slowly moved west, making maps and observations as they went, and by October had reached Albuquerque. By the following February they were in the valley of the Colorado River, where they made their camp 131.

Our plate shows this camp, with a band of Mojave Indians visiting it. These Indians were "lively and good-humored, cheerfully answering our questions regarding the way before us." It is possible to pick out camp 131 on the accompanying map, just to the south of the

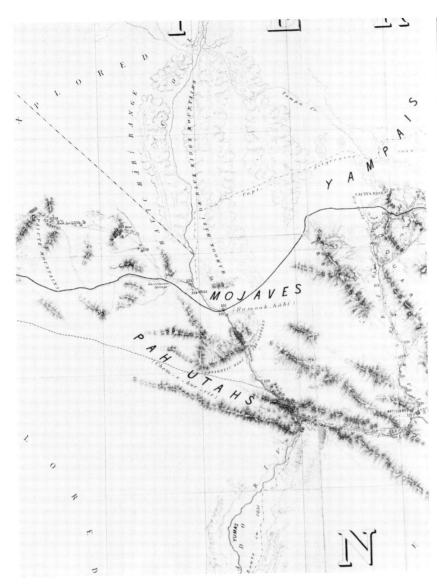

Plate 5.9a. Detail from the map accompanying the report cited at Plate 5.9.

"O" in "Mojaves," at the point where the expedition recommended that the railroad cross the Colorado River. The view was evidently taken from the east, looking across the camp to the river and then to the Havic Range.

It is perfectly easy to identify this point on the modern USGS map, about ten miles south of the town of Needles. In fact, all these railroad reports contain material that reveals a great deal about local history. One can use the maps, for instance, to track on the ground the routes that the explorers took and the camps they made. Then the texts can be brought in, to analyze the early human occupation of the region, as well as local flora and fauna.

Source: Reports of Explorations and Surveys, *vol. III (Washington, D.C.: 1856).*

Plate 5.10. "Mission Church of San Xavier del Bac," from the *Reports of Explorations and Surveys . . . 1853–1856* (vol. 7, Washington, 1857).

Plate 5.10a. Detail from Map no. 2, "From the Pimas Villages to Fort Fillmore," from *Explorations and Surveys made under the direction of the Hon. Jefferson Davis, Secretary of War* (Washington, 1857).

Meeting of Anglos and Hispanics in the Southwest

Down toward the 32nd parallel, the railroad survey was conducted by Lt. John G. Parke, whose report was also printed among the government papers. He worked during February and March of 1854 through particularly difficult country. The local Indians were not inclined to be helpful and needed, according to Parke, "complete and effective chastisement." The maps deriving from this part of the survey were particularly well drawn, as our detail of the area round "S. Xavier del Bac" shows.

This, of course, was country long penetrated from the south by the Mexicans, and they had left some spectacular monuments. The report includes a superb lithograph of the mission church at San Xavier del Bac, viewed from the west. One wonders what feelings went through the heads of the Anglo artists, as they confronted these evidences of an alien but highly developed culture. The mission was marked on the accompanying map, together with various early routes, both of U.S. Army engineers and of the Mexicans. Today this once remote mission has become almost a suburb of Tucson.

Source: Reports of Explorations and Surveys . . . 1853–1856, *vol. vii (Washington, D.C.: 1857).*

Plate 5.11. *California Crossing, South Platte River,* oil painting by W. H. Jackson, 1867. The Thomas Gilcrease Institute of American History and Art, Tulsa, Oklahoma.

The Perils of the Oregon Trail

In the course of an extraordinarily long life, William H. Jackson (1843–1942) made a very great number of sketches, paintings, and photographs. This oil painting of the upper California Crossing, on the South Platte River near Julesburg, Colorado, derives from sketches that Jackson made when crossing it in 1866. His own account of the occasion runs as follows: "Arrived at the crossing place on the South Platte, some two or three miles below Jules-

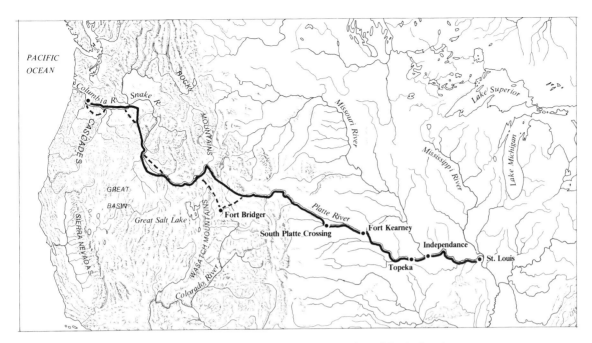

Plate 5.11a. Modern map showing the course of the Oregon Trail, and the California Crossing.

burg, July 24th. When we arrived, there were some three or four other trains preparing to cross. We uncoupled our wagons and put from twelve to eighteen yoke to each single wagon. Current so swift, it takes some of the smaller oxen off their feet in the deepest parts. Crossing at the same time were large bands of Sioux."

Travelers to Oregon and California had sooner or later to cross the South Platte, and this location eventually became the most frequented crossing. But it was difficult even when the river was low, and perilous when it was high; many oxen and wagons were swept away and lost. Jackson has drawn the fording from the south, looking north up Cambridge Hill. On the modern map, the fording would have been just to the south of the word "Pacific," where on the colored version of the map a certain amount of woodland shows in the center of the river, no doubt because here it was shallowest.

Jackson's painting is crowded with detail. On the left are two Indian tepees, with smoke pouring from their tops; the Indians themselves are standing around here. Just above the white horse, a wagon is taking the terrifying splash down into the river; others, with their great teams of oxen, are in midstream. On the right, two other wagons are waiting to go, with their teams hitched up, and on the extreme right a wagon awaits its turn, the occupants placidly sitting round the fire. In the center of the picture are two teams of oxen coming back for a fresh load. On the far bank, the wagons have re-formed into trains and are beginning to move off up Cambridge Hill. This is an extraordinary document, full of teaching potential.

Sources: Ralph Moody, The Old Trails of the West *(New York: 1963); and Irene Padden,* The Wake of the Prairie Schooner *(New York: 1943).*

Nineteenth-Century Landscape Views   139

Plate 5.12. *Hunter Mountain, Twilight;* oil on canvas (30½ × 54 in., 5.1983) by Sanford Robinson Gifford, 1866. Daniel J. Terra Collection, Terra Museum of American Art, Chicago.

An Image of the Expanding Frontier

Sanford Robinson Gifford (1823–1880) was exceptional among North American artists of the nineteenth century, for he was the son of a rich man, an ironmaster, and was able to travel widely in Europe. Among art historians, he is generally regarded as rather similar in style to Inness (Plate 5.7), but lacking the latter's masterly use of light.

Gifford himself felt this lack, and in 1865 went so far as to tour the newly opened Rocky Mountains, hoping for some fresh inspiration. But he could not adjust his style to their vast scale, and so returned to his Eastern subjects. His painting called *Hunter Mountain, Twilight,* completed in 1866, is a fine example of a work whose intellectual and emotional appeal does not depend upon any actual knowledge of place. It is instead the epitome of what it must have felt like to be settling the vast North American wilderness—*any* part of the wilderness—during the nineteenth century.

The house and outbuilding are dwarfed by the surrounding trees, just as the farmer and his cattle are symbolically dwarfed by the cleared stumps. Behind them all lower the wooded foothills and the great mountain, with no sign of human habitation on either of them, unless that very faint smoke far away at the foot of the mountain is intended to mark the cabin of another settler. This painting illustrates the way in which topographical views may sometimes make their effect as much by an appeal to the senses as to the intellect.

Source: James Thomas Flexner, History of American Painting, *3 vols. (New York: 1969–1970).*

CHAPTER SIX

Bird's-Eye Views of Towns and Cities

Gerald Danzer

As we saw in Chapter 1, the tradition of urban views and the principles of their construction, as well as the social and cultural context in which they were used, were well established by the time that urban places appeared in Anglo-America. In Chapter 3 (Plate 3.7) we saw an example of aerial perspective being used to delineate Savannah, carving itself into the wilderness of Georgia. There had been other views portraying New York City (1719–1721), Charleston, South Carolina (1737–1739), Boston (1743), and Philadelphia (1754).

Even before Independence, then, an audience existed on both sides of the Atlantic for views of colonial towns. After the Revolution the genre continued to thrive. As the nation pushed westward, each new place on the urban frontier became a candidate for a view "suitable for framing." The development of lithography in Europe, outlined in Chapter 5, made it possible to produce handsome colored prints at a modest cost. Lithographic prints, widely produced in North America, became the democratic art, and no subjects were more eagerly purchased than cityscapes and townscapes.

John Casper Wild was one European artist who sensed the opportunities available across the Atlantic. Born in Switzerland, he made a living as a landscape artist in Paris for nearly fifteen years. In 1831 he moved to Philadelphia and began drawing views of that city, but after a few years sensed that better fortune awaited him in the newer cities. He went to Cincinnati to begin a series of lithographs of the Queen City of the West, but again after a year or two had second thoughts, and returned to Philadelphia. He published a series of prints of individual streets and buildings, selling them in sets of four and planning to collect them later in a book. His set of four views from the State House steeple (1838) surveyed the entire city. Once again, however, he left Philadelphia to set up business in a western city, this time St. Louis. After a brief stay there, he started his last large project, planning to produce views of all the towns of the upper Mississippi valley.

The resulting prints show Wild's boundless energy and sense of enterprise. They were important ingredients in the success of this type of booster literature, when so many views were in effect commissioned by subscribers. John Reps, the leading authority on nineteenth-century views of North American cities, estimates that more than 2,400 separate places were the subject of prints. As a whole, these views constitute an impressive collection of documents illustrating the urban ideal, as well as an archive of details on the look of individual buildings, streets, and settlements.

Each town had its own distinctive look and every artist had a unique way of drawing it. Both the particular form of the individual town at any given time, and the specific statement by the artist have value as historical sources. As far as the drawing was accurate, it has immense utility for the urban or social historian. In so far as the view was a personal vision colored by the hopes and dreams of its artist or patron, the value of the source moves to the realms of cultural and intellectual history. The great challenge to any historian using an urban view, or an urban plan for that matter, is to separate fact from fiction. The difficulties entailed in this process have indeed often led historians to avoid using the views as primary sources, preferring instead to employ them as illustrations rather than documents. Yet we have no better source to tell us what a particular town or city looked like, as a whole community, at a particular date. Nor do we have any other source that is so geographically widespread, so representative of the society as a whole, or so close to the actual urban form, that also tells us how North Americans wanted their communities to appear. As scale models or as grand visions, bird's-eye views furnish a striking perspective on their civilization, especially in the nineteenth century, the heyday of the medium.

Local historians have long used reproductions of these lithographs to supplement or to illustrate their general accounts. Few writers, however, have started with the views as basic sources from which to reconstruct the life of a community in a particular period. It is rare to find a published account that uses one of these panoramas to describe the city as a functioning entity, although that was a major purpose for which they were made.

A selection from the thousands of extant views provides extensive documentation for the urban vision in nineteenth-century North America. In looking at a sampling of the urban views, one is struck by how little they change, region to region and decade to decade. Each print seems to reiterate a basic refrain: cities everywhere all seem caught in the process of expanding. Streets and roads push the urban area into the surrounding countryside, and one can almost hear the hum of commerce in crowded but orderly harbors and depots. Railroads, trains and steamboats, coming and going across the scene, serve as reminders of how this particular city is linked by iron bands and blue waters to a whole system of urban places across the continent.

The sun is always shining, the season is usually early summer, and the gentle breeze blows just enough to unfurl the flags of freedom and to highlight the smoke sent out as an emblem of progress from the steam engines of factories, locomotives, and ships (Plate 6.8). As a group the urban views sing the national anthem of peace and prosperity, of movement and openness, of calm and order, and of destinies to be fulfilled. The cultural baggage carried by the images immediately makes them suspect, yet the few studies made of particular buildings shown in the views have, on the whole, been a tribute to the accuracy of their details. David Ruell (see the bibliography at the end of this chapter) concludes his study of New Hampshire prints made between 1875 and 1899 with the observation that, "as a general rule, unless conflicting evidence surfaces, the bird's-eye views can be trusted as correct and complete representations of their subjects."

It might be best, therefore, to think of the views as flattering urban portraits that followed certain rules of perspective, setting, mood, and overall style. Because they were sold as commercial ventures to the townspeople, they had to look accurate to be convincing, but not so honest as to reveal the problems and imperfections of their subjects. That is why the distance of the bird's-eye viewpoint and the perspective chosen by the artists were so advantageous. They gave the artist license to select

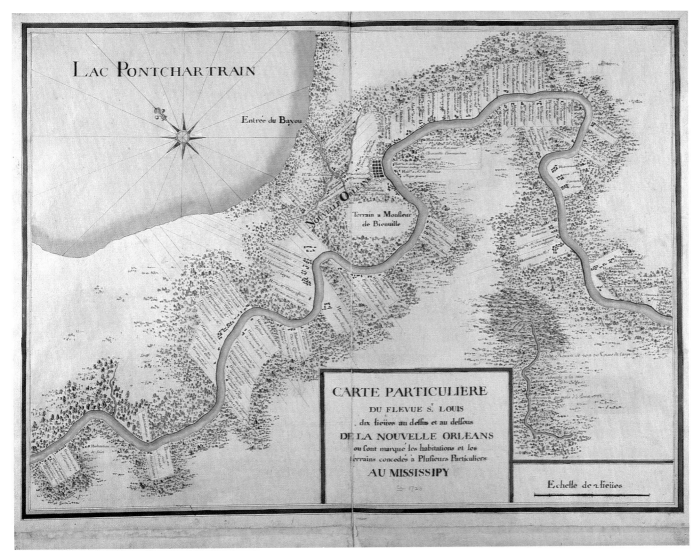

1. The Mississippi River in the neighborhood of New Orleans; manuscript map from the "Cartes Marines," (c. 1720). The "Cartes Marines," preserved at the Newberry Library, are a collection of about one hundred manuscript maps, showing French overseas possessions in the early eighteenth century.

2. LANDSAT image of the same area as Plate 1. From *Man On Earth* (p. 32) by Charles Sheffield; reprinted by permission of Sidgwick and Jackson and the Macmillan Publishing Company. Copyright © 1983 by Charles Sheffield.

This image does not cover exactly the same area as Color Plate 1, but it is oriented in the same direction, and by picking out the distinctive southward bulge of New Orleans (center on the manuscript, right on the LANDSAT image) the reader can identify other common features. Note particularly the prevalence of the long-lot pattern in the modern landholding system.

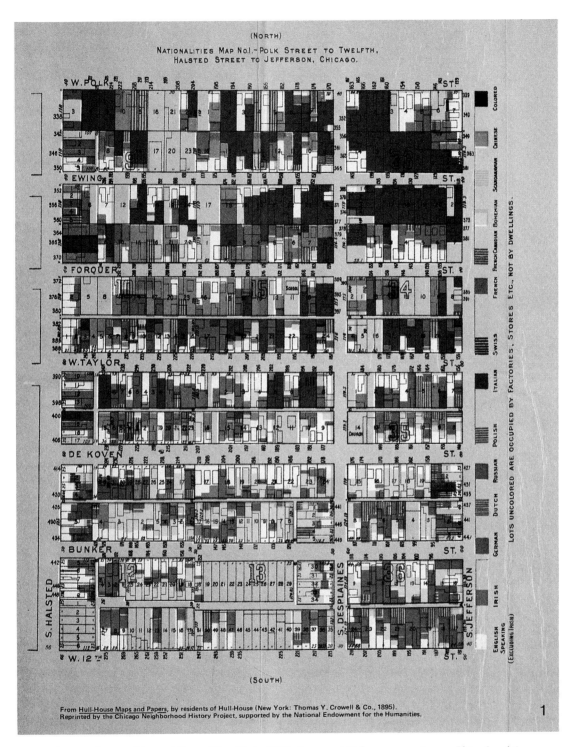

From Hull-House Maps and Papers, by residents of Hull-House (New York: Thomas Y. Crowell & Co., 1895).
Reprinted by the Chicago Neighborhood History Project, supported by the National Endowment for the Humanities.

3. Plate from *Hull-House Maps and Papers* (New York, 1895). See page 167 for a commentary on Plates 3 and 4.

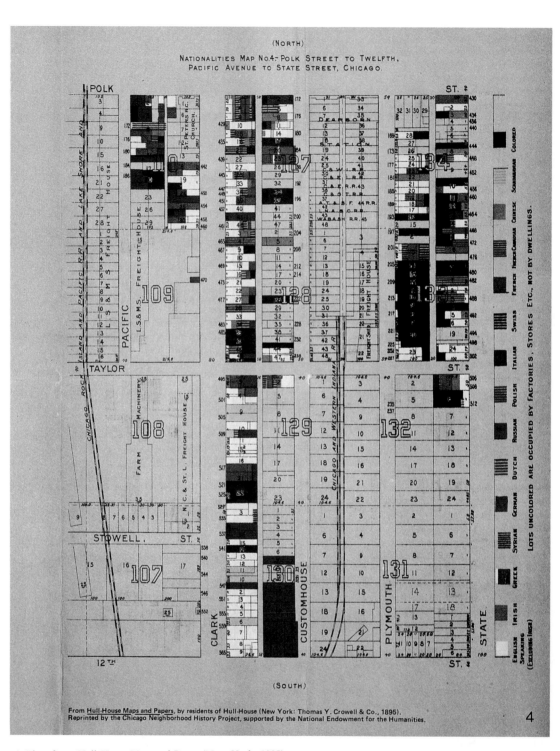

4. Plate from *Hull-House Maps and Papers* (New York, 1895).

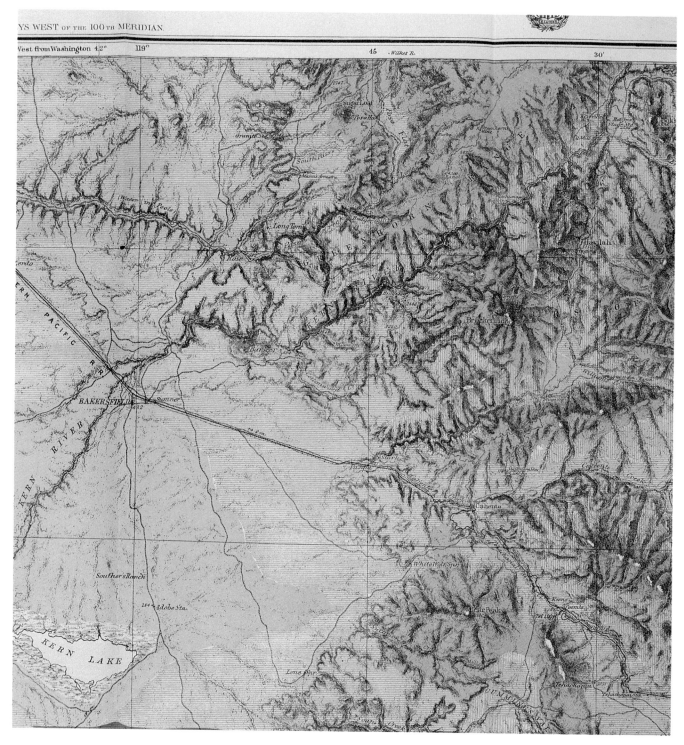

5. Detail from sheet 37 of the Wheeler Survey. See page 247 for a commentary on this map.

6. Cover of the *Official Road Map of New Mexico* (1936). New Mexico State Highway Department, Santa Fe. See page 263 for a commentary on this type of map.

42°00'—

41°45'—

7. LANDSAT image of the Chicago area. Illinois State Geological Survey and Northern Illinois University, DeKalb. See page 291 for the commentary on this plate.

8. Aerial view of Fort Union, New Mexico. William Garnett. See page 307 for the commentary on this plate.

and modify details. For example, houses in the best part of town might display fine grillwork and precisely rendered windows, while poorer neighborhoods were filled with generic structures having few distinguishing features. The artist's attempt to get all the important buildings into the picture, and the frank promotional function of the prints, introduced elements of distortion that tended to mask a town's shortcomings. Photographs could seldom be so flattering.

Almost every view conformed to some general conventions. Each artist had his own way of going about the business of depicting a town, but most followed a basic procedure. First the general street plan was sketched on paper. Then this was modified by stretching and widening the streets so that drawings of individual buildings, usually quite accurate, could be fitted onto the street pattern at the proper locations. The focus of the view, of course, was ordinarily the central business district. We can call this the focal area or middle ground. Surrounding it were the other standard elements of composition: foreground, background, and surroundings, which make transitions from the focal area to the sides of the view. An ornamental frame usually closed in the whole composition, providing space for a title, key, and written description, or additional views and vignettes.

Each element in the composition had a different function and often carried a different level of accuracy. The foreground played a symbolic role, with great care lavished on the details of a bridge, a ship, a tree, a rocky outcrop, or an important building. The foreground is thus the best place to look for personal statements by the artist. If people were included in the picture, this is where they ordinarily would be seen. The middle ground was usually the most accurately drawn part of the scene. In the background the images were more generalized. The regions surrounding the town were often drawn without details, except for hints here and there of the future expansion of

the urban area and suggestions about the natural environment of the place. The horizon between the background and the sky was usually a line parallel to the top of the frame. This horizontal emphasis stressed the peaceful mood of the place, providing a sense of openness and boundlessness, as on Plate 6.2. An 1890 print of Colorado Springs, labeled "Pikes Peak Panorama," even blends the snowcapped peak into a horizontal arrangement.

The sky itself was often reduced to a thin ribbon across the top of the view, a result of the oblique angle at which the artist, imaginatively perched on a cloud, looked down at the city. One result of this perspective is that the city seems to dominate the landscape. A vast sky might have the effect of shrinking the civic achievement to that of a speck in the cosmos. An emphasis on urban mastery distinguishes the bird's-eye view from the landscapes and rural scenes shown in Chapter 5. Shakespeare's "majestical roof fretted with golden fire" was often reduced to a narrow sky, framing the attainments of civilization recorded in buildings, streets, and public works.

The publication of the monumental *Views and Viewmakers of Urban America* by John W. Reps in 1984 has opened the bird's-eye view to scholarly inspection as a primary source. Its subtitle, *Lithographs of Towns and Cities in the United States and Canada, Notes on the Artists and Publishers, and a Union Catalog of their Work, 1825–1925,* hints at the richness of the volume's resources. It reproduces over a hundred of the views, provides biographies of the fifty-one major viewmakers, and includes an extensive bibliography. In this book and others, Reps does a masterful job of putting particular views in their cultural and historical context. Following his techniques, the problem for the teacher or researcher is not so much one of locating views of particular urban areas as of placing them in their cultural context and developing an appropriate mode of analysis.

Plates 6.1 to 6.4 analyze a single very rich print of Chicago. Its artist, James T.

Palmatary, arrived in Chicago early in 1856, when the city was just emerging as a major transportation center. The Illinois Central and Michigan Central railroads had just reached the mouth of the Chicago River, and the Illinois and Michigan Canal had been in operation for eight years. Chicago had thus become the hub of a transport system stretching down to the Gulf of Mexico and out across the Great Lakes to the Atlantic.

Palmatary seems to have gone at once to the editor of the *Daily Democratic Press,* with a proposition to publish a bird's-eye view of this emergent metropolis. According to an account printed in that paper, Palmatary had "sketched and lithographed" over fifty other cities, and now proposed a print measuring 52 by 34 inches of Chicago, for which advance subscribers would pay ten dollars. This was a large sum, but it was a large print, and in fact it had to be enlarged during manufacture, to take in all the necessary buildings. The work went forward for five months, and then the editor declared himself satisfied with the view, "better than we had supposed could be reduced to paper." Some improvements and new buildings, "so far as they are actually determined upon or commenced, are represented as they will appear when completed;" thus Palmatary could justify a date of 1857 in the title even though he actually drew the city in 1856.

The work was sent to Philadelphia to be lithographed, the editor concluding his promotional piece with the words: "All who can derive advantages from the possession of so fine a view of the Garden City should lose no time in subscribing for the work with Messrs. Braunhold & Sonne," the Philadelphia lithographers. The view was thus produced and purchased "for advantages;" it was commercial art and not art for art's sake. By November 1856 the editor could report that he had seen a "proof impression" of part of the view, and that this was on display in his offices so that "those who

are anxious to know that their buildings are rightly located can call to see it." A month later another sheet arrived, showing another part of the city: "as far as we know every house in the section of the city represented is faithfully drawn." Perhaps the comment on individual houses was designed as an appeal to individual residents to purchase the view, though its large size (eventually 44 × 80 inches) would have limited its use in most homes.

We do not know if the venture was a success in business terms. However, its promotion was interestingly encouraged by the production of a smaller print, about 22 by 13 inches, showing Chicago in 1820. This smaller view, drawn by another artist and lithographed in Chicago, was given away as a "premium to Palmatary's View of Chicago in 1857." It showed a barren site with only a small fort, a trading post, and a few cabins denting the vast expanse of treeless prairie, using the bird's-eye perspective to show a site waiting to be developed. The premium thus pointed out the progress of civilization over the past generation. It served as a witness to the transformation celebrated on the larger lithograph, and placed the 1857 view in a temporal context. Packaging the two views together was not only a sound business move; it was also a striking commentary, providing an insight into the cultural context of these views in the Age of Improvement. The juxtaposition of a frontier scene with the emergence of Chicago as a transportation hub demonstrated a sensitivity to time, change, and milestone events that speaks clearly to later historians.

What we may also hear, in listening to these remarkable pieces of evidence, is not only an account of how things actually were in any given town, but also how the artist and patrons wanted them to be. They are sources that allow us to probe the nature of a particular urban place both as a physical artifact and as a functioning entity at one point in time.

SOURCES FOR BIRD'S–EYE VIEWS

Beckman, Thomas. *Milwaukee Illustrated: Panoramic and Bird's-Eye Views of a Midwestern Metropolis, 1844–1908*. Milwaukee: 1978. An exhibition catalog that traces the pictorial record for one city.

Bumgardner, Georgia B. "Graphic Arts: Seventeenth–Nineteenth Century." In *Arts in America: A Bibliography,* vol. II, section K, ed. Bernard Karpel. Washington, D.C.: 1979. A useful collection of suggested readings with a special section on urban views.

Clay, Grady. *Close-Up: How to Read the American City*. Chicago: 1980. A handbook designed for observing the contemporary scene, but equally useful for looking at old views with a discerning eye.

Hales, Peter Bacon. *Silver Cities: The Photography of American Urbanization, 1839–1915*. Philadelphia: 1984. This attempt to trace changing "American attitudes toward the city as revealed in an evolving medium" provides as well some ideas for ways of looking at lithographs.

Hébert, John R., ed. *Panoramic Views of Anglo-American Cities*. Washington, D.C.: 1974. A checklist of bird's-eye views in the Library of Congress.

Historic Urban Plans. *Historic City Plans and Views*. Ithaca, N.Y.: 1985. A catalog of reproductions featuring North American, European, and some non-Western cities. These reproductions are a key resource for the study and teaching of urban history and should be found in any major library.

Mumford, Lewis. *The City in History*. New York: 1961. This classic work is useful for studying urban prints because of its approach and the variety of its source materials.

Rand McNally and Company. *Bird's-Eye Views and Guide to Chicago*. Chicago: 1898. Reprinted in Frank A. Randall. *History of the Development of Building Construction in Chicago*. Urbana, Ill.: 1949, pps. 151–217. The plates provide a block-by-block portrayal of the city's central business district and an accompanying text comments on all the major buildings.

Reps, John W. *Cities on Stone: Nineteenth Century Lithographic Images of the Urban West*. Fort Worth: 1976. This book remains useful as an introduction to the subject.

————. *Panoramas of Promise: Pacific Northwest Cities and Towns on Nineteenth-Century Lithographs*. Pullman, Wash.: 1984. A regional study with fifty full-page illustrations and several suggestions on how to use the views as documents.

Saint Louis Illustrated: Nineteenth-Century Engravings and Lithographs of a Mississippi River Metropolis. Columbia, Mo.: 1989. Discusses over 100 views, often using quotations from contemporary visitors.

————. *Views and Viewmakers of Urban America: Lithographs of Towns and Cities in the United States and Canada, Notes on the Artists and Publishers, and a Union Catalog of their Work, 1825–1925*. Columbia, Mo.: 1984. The essential reference volume on the topic. It reproduces over a hundred of the views, provides biographies of the fifty-one major viewmakers as well as an extensive bibliography, and includes two especially useful chapters in its introductory section: "Drawing the Views: The Artist and His Subject," and "Lithographic City Views: Reliable Records of the Urban Past."

Ruell, David. "The Bird's-Eye Views of New Hampshire: 1875–1899." *Historical New Hampshire* XXXVIII:1 (1983), pp. 1–67. Contains many citations to contemporary newspapers.

Stokes, I. N. Phelps, and Daniel Haskell. *American Historical Prints: Early Views of American Cities*. New York: 1932. Both an introduction to the extensive collection in the New York Public Library and helpful notes on individual prints.

Stokes, I. N. Phelps. *Iconography of Manhattan Island, 1498–1909*. 6 vols. New York: 1915–1928. Unequalled in scope, these magnificent volumes use the bird's-eye views and other sources to describe the changing cityscape.

Tunnard, Christopher, and Henry Hope Reed. *American Skyline*. New York: 1955. Seven tables at the conclusion of the volume describe the various aspects of the city in each particular era of North American history.

Plate 6.1. Detail from *View of
Chicago* by James Palmatary
(Philadelphia, 1857). The
Chicago Historical Society.

The Emergence of Chicago as a Walking City, 1857

Palmatary's large view of Chicago is a splendid portrait of a North American town at the height of its development as a walking city. Two years after its publication Chicago would receive its first streetcar line, and the transformation into a streetcar metropolis would begin. The city was at this very moment losing the last vestiges of a frontier settlement, for the remnants of Fort Dearborn were being removed just as the artist was making his drawing.

One must be aware of just when a bird's-eye view was drawn, to be able to "read" it accurately. Palmatary left no traces of the fort or the frontier origins of the city; instead, he emphasized the city's new role as a commercial and transportation center. Ships, waterways, locomotives, and railroad tracks animate the scene and provide a definition of the city. It comes as something of a surprise to learn that the Illinois and Michigan Canal was completed only eight years before this view, and that railroad tracks did not reach the city until that same year, 1848. The impressive complex of terminals, warehouses, and grain elevators on the south bank at the mouth of the river was in the process of construction when Palmatary was making his sketches, some of the buildings occupying the very site of the Fort Dearborn military reservation that had been sold by the federal government to the Illinois Central Railroad. In fact, Palmatary was portraying a city that was as novel to native Chicagoans as it would have been to visitors, and this may explain why the print was published in such a large and elaborate format.

The newness and vitality caught by the artist may also explain an unusual feature of Palmatary's view that seems like an attempt to plant some roots in the city's past, or to connect it to national events. This is the curious inclusion of a group of soldiers or militia marching in Lake Park just south of the site of the old fort. This feature is lightly drawn and must be pointed out to be observed, even on the original print. The figures are drawn out of scale, about the same size as the trees in the park or the buildings on Michigan Avenue. The inclusion of such a feature perhaps serves to remind us of an era that has just closed, Chicago's frontier period. The soldiers on parade forcefully point out that the bird's-eye views were cultural statements as well as pictorial references. The challenge placed before the student is to see both dimensions.

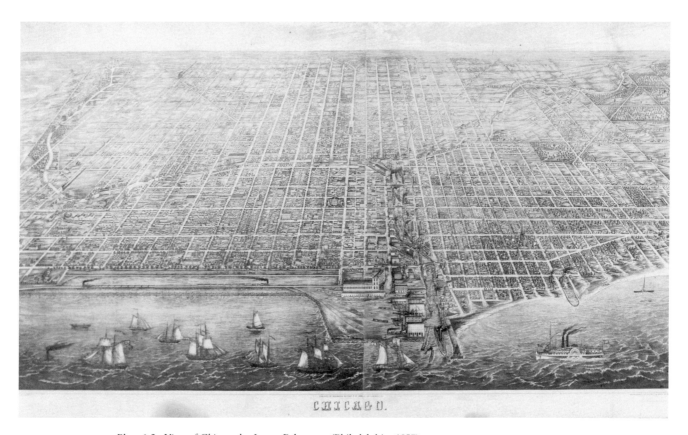

CHICAGO.

Plate 6.2. *View of Chicago,* by James Palmatary (Philadelphia, 1857).

Palmatary's View of Chicago in 1857, Seen as a Whole

The composition consists of several basic elements or areas. Start with the sky, and note how this artist, like many others, has suggested a flat horizon, to give a peaceful frame to the composition. Next, search the foreground for evidence of symbolic statements; this would be Lake Michigan in the present example. The variety of lake vessels surely suggests something about the diverse cargoes coming into the port. On the actual view, it would be easier to observe that the schooner approaching the harbor is the *Dean Richmond,* which was the first vessel to sail from Chicago to England. The artist prophetically included its return to the home port a year later, establishing Chicago as a world market.

The third element to note is the differing accuracy of various parts of the view. The center, where the Chicago River flows through the business district and out to the warehouses and docks, is very accurately drawn. The areas to the north (right), west, and south show a much simplified landscape where cultural features are only suggested. Usually these transition areas show signs of the city expanding out into the hinterland. The heart of downtown is the most carefully drawn, with the perspective and the distances carefully adjusted to provide good views of most of the important buildings. In many city views these are identified by a key located below the picture or in an elaborate frame around the print.

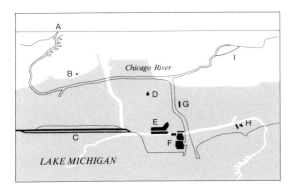

Plate 6.2a. Diagram of
Palmatary's *View of Chicago.*

Palmatary's choice of vantage point employs the Chicago River and Lake Michigan as dividing lines to delineate districts within the city. The focus of the view is on the main course of the river, which divides the city into a north and a south side. The north and south branches of the river create a west side that at this time would have been considered part of the surroundings, not yet a vital part of the city. Lake Michigan provided a sharp boundary for the city to the east; thus the viewpoint of the observer was looking due west down the main stem of the Chicago River.

The street pattern provides another important orientation to the view, most of the streets being labeled on the print. The density of the street pattern is an important clue to the density of settlement in the outlying areas. Careful observation of the street pattern can also locate boundaries between subdivisions, diagonal streets that often followed old Indian trails in spite of the gridiron urban plan, and the ways in which the new railroad tracks were accommodated in the layout of the town. Even the river has been channelized and the lakeshore filled in, to parallel the gridiron system of streets and avenues.

In studying a very large view like this one, it is essential to identify a number of sites that a modern student could be assumed to know. Here, they are lettered on the accompanying sketch. At "A" is the great trestle of the Illinois Central railroad, set out in Lake Michigan to reach around the built-up city. At the mouth of the river, "B," the grain elevators and railroad terminals cluster together, ready to reload grain into the lake schooners. "C" marks the waterworks, at the foot of Chicago Avenue. The present Water Tower and pumping station replaced these facilities in 1869 at the same location.

The Galena and Chicago Union Railroad Terminal ("D") on the north bank of the Chicago River brought produce and passengers from the region west of the city. The tracks, later part of the Chicago and Northwestern Railway, are still partially intact underneath the modern buildings. In 1856, Charles Hull was building a country house on the edge of the built-up area southwest of the central business district. The famous house on Halsted (Hull-House, scene of the labors of Jane Addams) did not appear on this view (site at "F"). Away to the southwest ("E") appears Bridgeport, terminus of the Illinois and Michigan Canal, which had been a major cause of Chicago's extraordinary expansion.

Most of the city pictured here was destroyed in the Chicago Fire of 1871. Even the lakeside has not remained constant, the city of today extending several blocks into the lake (see Plate 4.4). Still, enough remains of the general plan, together with some specific landmarks, for a plan like this to provide an excellent exercise for students.

Plate 6.3. Detail from Palmatary's *View of Chicago*.

The Use of Detail in Studying Bird's-Eye Views

Wall-sized bird's-eye views were often designed by the artist not only to provide an overall impression of the urban form, but also to furnish insights into the city at work. The detail from Palmatary's view shown here presents the central point of interest: the harbor at the mouth of the river where railroads from every direction brought goods to be exchanged with the cargoes of the lake boats.

The grand depot of the Illinois Central and Michigan Central railroads dominates the scene. In actual fact, the structure was not as tall as the Sturges and Buckingham Elevators, which were, for many years, the largest structures in the city. The depot was under construction when Palmatary sketched the view. By turning it on a slight angle he was able to suggest its elaborate facade; such attention might stimulate patronage by railroad officials. At any event, there is here no question that the age of rails had arrived and that the main railroad station had become the focal point of the city.

The Marine Hospital with the colonnaded facade and ample yard was the only remnant of the old Fort Dearborn military reservation that had occupied the site only a few months earlier. On the north bank of the river, to the right in the view, lumber from the northern reaches of Lake Michigan is stacked in neat piles. The government pier is shown effectively performing its function of keeping the harbor free of the shifting sands. Upstream, beyond the Rush Street swing bridge, is a building that looked like a castle. This was the grain house of the Galena and Chicago Union Railroad, which received wheat from farms west and north of the city. The facilities on the other side of the river serviced railroads that extended to the south and east.

Thus although the focal point of the view centers on the shallow mouth of a sluggish prairie stream, the age of invention and enterprise has brought to the site products from many parts of the region and perhaps even from the world beyond. In that sense the details of the city at work suggest an even larger canvas and a global perspective.

Plate 6.4a. (*above*) and b
(*opposite*). Close detail from
Palmatary's *View of Chicago*.

The Use of Fine Detail in City Views

It is interesting to see how far a student can push some of the details on a well-crafted bird's-eye view. The enlarged details on this plate push Palmatary's view of the city to the limits of our ability to photograph them. Each detail can be located on Plate 6.3, from which these are what might be called "fine details." The structure and function of the Rush Street bridge are clearly visible in the enlargement. A team of horses is shown approaching the structure from the north, and a group of pedestrians has congregated at its southern edge. Perhaps the bridge has just swung to its closed position after allowing a ship to pass. The rows of buildings along the avenues illustrate another aspect of the walking city: the mixture of land uses, building types, and economic classes on the same block. In this detail, only two blocks south of the bridge, the variety of the urban scene is apparent: modest cottages, large houses, commercial buildings, sheds, yards,

gardens, and vacant lots are all mixed together. Fortunately the same kind of scene can be documented in photographs taken by Alexander Hesler at approximately the same time. Some of these are reproduced in *Chicago: Growth of a Metropolis* (Chicago, 1969) by Harold M. Mayer and Richard C. Wade, pages 9 and 38, and on the foldout following page 116. The comparison of contemporary photographs with the details of the bird's-eye views is a good way to check the accuracy of the print.

In the present case, too, this mixture of land uses raises many interesting questions. Was society as a whole better served by a residential pattern where rich and poor rubbed shoulders at home? What implications might there be for environmental concerns when factories were cheek by jowl with homes? How might the mixture of land uses have influenced the nature and power of municipal government in the walking city?

Plate 6.5. *Los Angeles,* by Kuchel and Dresel, 1857.

The Hispanic Origins of a Great Metropolis

A major question facing students of North American urban history is the extent to which the origins of modern cities lie in the preindustrial past. How much of an impact did the preindustrial town have on the later metropolis? Do the formative years for modern communities really arrive only in the mid- or late-nineteenth century, after the appearance of technological advances in transportation, building construction, and industrial production? Does this view of Los Angeles in 1857 have an integral place in the development of the city? Or is it only a quaint heritage, at most a kind of prologue that fills up the chronological space before the real story of Los Angeles begins?

Questions like these often emerge when one looks at the earliest views of North American cities. This idealized portrait of an Hispanic town, though contemporary with Palmatary's print of Chicago, seems to be of a much different order and to take a much different place in the interpretation of a city's past. The view of Chicago, emphasizing the dynamism of commerce driven by the railroads and the lakeboats, seems directly connected to the Chicago that we know today; but without a caption, even careful observers would have difficulty in identifying this place as Los Angeles.

The story of the print itself, however, helps put the scene into perspective. Charles C. Kuchel and Emil Dresel formed a company in 1855 that produced in the short span of four years at least 50 views of towns in the West. Artists and printers flocked to California along with the miners and other fortune seekers. In 1849 the first view of San Francisco was pub-lished in Philadelphia. By the time Kuchel and Dresel arrived on the scene to record the mining towns in their boom period, San Francisco had become a cultural center and nearly all of their prints were lithographed there.

Los Angeles also felt the impact of the gold rush. Founded as a mission in 1781, the same year as the Battle of Yorktown, the *pueblo* entered a new stage in 1833, when the Mexican Republic opened the mission lands to ranching interests. Los Angeles as a small cattle center was thus put in an advantageous position to supply the gold towns further north with supplies during the 1850s. This is the Los Angeles that Kuchel and Dresel drew in 1857.

The artist has taken a position above First Street, looking north; Main Street is on the left, Los Angeles Street on the right. Arcaded streets, adobe walls, flat roofs, and enclosed courtyards dominate the cityscape, forcefully reminding modern viewers that a major aspect of the national heritage is rooted in Hispanic traditions. Only here and there do Yankee buildings with pitched roofs rise above the horizontal lines of the settlement. Perhaps these interlopers, however, were the seeds of the modern city, especially after the "great drought" of 1863–1864 wiped out the cattle industry. Or does the imprint of the Hispanic town, with a lineage reaching back to the Law of the Indies proclaimed by Philip II on July 4th, 1573, have a place in the modern metropolis? This quaint view thus opens up a major question in American urban history, and challenges the viewer to reflect on the very essence of the American character.

THE CITY OF NEW YORK.

NEW YORK, PUBLISHED BY CURRIER & IVES, 115 NASSAU ST.

Plate 6.6. *New York,* by
Currier and Ives (New York,
1867).

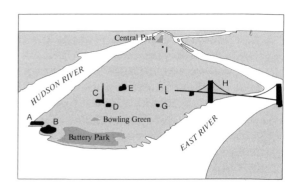

Plate 6.6a. Diagram of Plate 6.6.

The City of New York in 1876:
A Centennial Celebration

In 1834, exactly a century after Gordon's view of Savannah (Plate 3.7), Nathaniel Currier arrived in New York to seek his fortune. A year later he issued his first two prints, both news-type urban scenes of a great fire that destroyed part of the city. His firm, known after 1857 as Currier and Ives, issued over four thousand popular lithographs during the nineteenth century. About one fourth of this huge output could be classified as urban scenes, and of these several dozen used the bird's-eye perspective.

No one caught the spirit of the United States, the thrust of its aspirations, or the character of its memories, more graphically than the artists who worked for Currier and Ives. This cityscape therefore carries two great advantages as a document of the nation's history: the most celebrated imprint in North American lithography and a view of its largest city at the time of the centennial of independence. It was a period in which North Americans placed on exhibit for the whole world to see what John Brinckerhoff Jackson has called "their delight in organizing space and time and labor, their eagerness to acquire new ideas, their abundant creativity." (*American Space: The Centennial Years, 1865–1876,* pp. 239–40).

Few artifacts document these characteristics more vividly than this bold print. The city itself is a massive arrowhead thrust into the boundless interior of the continent. Suburbs and dependent facilities crowd the rivers surrounding Manhattan. Six or seven years hence, the giant towers of the Brooklyn Bridge (H) would create webs of cables to connect the heart of the metropolis to its sister city across the East River; on the print, though, the great bridge appears as an already functioning structure. Organized energy animates the placid waters and the broad thoroughfares, as well as the bridge. Ships and boats, carriages and wagons, street cars and pedestrians seem to be coming and going. As a whole their individual movements are balanced one with another and blended into a symphonic unity.

Dozens of identifying captions at the bottom of the print engage the viewer in a tour of the city. Starting with Castle Garden (B) and Battery Park, the labels lead the viewer into the urban fabric, past the Bowling Green and Trinity Church (C) to the Stock Exchange (D), City Hall (E), and on to the distant Saint Patrick's Cathedral (I) and Central Park. Lesser places are also noted: the Iron Steamboat Company's Pier (A), the Turner Building, the Germania Savings Bank (G), Jones' Wood, and the Beekman Street Shot Tower (F). These details suggest that the view was designed as much for close study at a table as for hanging on a wall. In either case, as an impression of the city or as a detailed introduction to the urban structure, the print is a celebration of North American progress at the mark of one hundred years of independence.

THE CITY OF SAN FRANCISCO.

BIRDS EYE VIEW FROM THE BAY LOOKING SOUTH-WEST.

Plate 6.7. *San Francisco,* by
Currier and Ives (New York,
1878).

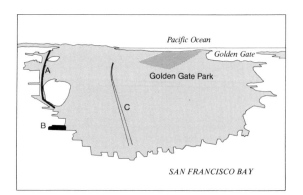

Plate 6.7a. Diagram of Plate 6.7.

San Francisco in 1878:
The Metropolis of the West

The similarity between the Currier and Ives print of San Francisco in 1878 and their view of New York City two years earlier is so striking that they seem to belong together as a pair—bookends holding the volumes of the United States together, furnishing vistas into the reaches of the continent and ports to connect the nation's shores with the wider world. Even the geography of this view seems contrived to resemble that of New York City. The overall impression is that the sea is at our backs and the horizon reaches into the interior. Yet the directions are in fact exactly the opposite. The subtitle explains that the artist is looking down on the city as from a balloon high above the Bay, westward to the Pacific Ocean. The arm of the sea to the right is the Golden Gate reaching outward to the vast ocean, not, as it may appear, an estuary reaching inland like the Hudson River.

The topography of San Francisco, like its location, seems to be reworked by artistic licence to resemble the great metropolis on the opposite end of the continent. The hills are made plain and the valleys exalted to announce the American century, an epoch perhaps opened when the iron rails, crossing the mudflats to the left (A), reached across the continent to link up with the tracks conspicuous along the New Jersey shore, to the left in the New York view.

Market Street (C) in San Francisco advances boldly into the city, cleaving it into two sections. The thoroughfare seems to be an echo of Broadway in New York. In place of the latter's Castle Garden, the California city features the Ferry Building (B). The ships in the two views are similar, interchangeable parts. At the far inland extent of both cities large parks clearly mark the landscape, Golden Gate Park symmetrically balancing Central Park. Church steeples define both skylines; the only thing lacking in the West was a great bridge.

The prints are about the same size, issued by the same publisher, addressed to largely the same audience, rendered in the same style, painted in the same colors, and probably done by the same artist. In 1888 James Bryce concluded that California, more than any other part of the Union, was a country unto itself with San Francisco as its capital. In the context of this Currier and Ives lithograph, published for the nation in New York City, the opposite seems to be true. The metropolis of the West Coast, the print seems to say, is the counterpoint to the one on the Atlantic. Both cities came from the same mold, so that although the views present the particular shape and texture of each city, they do so in the context of a national message and a continental theme. The national thrust, so evident in this view, reminds us to look for a similar stance in other nineteenth-century bird's-eye views.

PERSPECTIVE MAP OF

FORT WORTH, TEX.

1891

Plate 6.8. *Perspective Map of Forth Worth, Tex.,* by Henry Wellge, 1891.

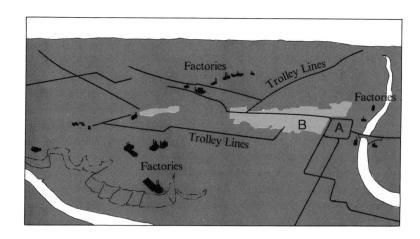

Plate 6.8a. Diagram of Plate 6.8.

The Apotheosis of the Bird's-Eye View:
Fort Worth, 1891

By the time Henry Wellge produced this splendid print of Fort Worth in 1891 the tradition of the lithographed view was drawing to a close. The artist, an immigrant from Germany, would continue drawing cities up to 1910, but the bulk of his work—more than 150 urban scenes in twenty-six states—preceded his second perspective of Fort Worth. After this he would issue only six more prints.

In a sense Wellge's second portrayal of Fort Worth may therefore be considered the culmination of the tradition. No view was produced with finer artistic standards or better use of color. In this case the print is called a "toned lithograph" because the colors were applied by hand. The inset of the Hurley Office Building not only indicates the patron of the project, but provides a clue to the interpretation of the city in 1891 as well.

Note that the building is seven stories high, "about as high as a building ought to go," in the words of the song. A seven-story building indicates a new technology and a new shape for the city. Most certainly we can expect to find streetcar tracks holding this city together, and a close inspection shows that this is the case. Furthermore, note how the streetcars have sorted the city out into distinct districts. One can almost draw boundaries round the central business district (B), the government square (A), the transportation corridors, several industrial districts, and neighborhoods of various types, each with its own schools and churches. It is striking how the smokestacks are gathered together like trees in groves.

One might even speculate on the rela-tionship between the emergence of the streetcar city and the passing of the vogue for panoramic maps of the nation's cities. The streetcars pushed urban boundaries so far out from the center that it was no longer possible to portray the whole metropolitan area in one view. A nascent city like Fort Worth in 1891 is an exception. Similar views of New York, Philadelphia, Boston, or Chicago would have had to be gigantic in size to accommodate the streetcar sprawl. Wellge's last prints were of places like Billings, Montana; Chippewa Falls, Wisconsin; and Duluth, Minnesota. His attempt to do a "Panoramic View of Greater Milwaukee" in 1907 marks the end of the genre. The original ink drawing was so large that it was not published as a lithograph. Instead it was photographed to reduce it to a manageable size. Even then, a careful viewer has noted that the original drawing was left unfinished, with many blocks indicated in outline only.

Perhaps the focus of a citizen's interest became more narrowly focused on the local neighborhood, the place of work, and the central business district. If a city like Fort Worth would keep growing in every direction, it might get too big in a physical sense for one print, and might also expand too far in a cultural or psychological sense as well. Did one's sense of the city really extend to all of the hundred or so community areas into which the early sociologists divided Chicago? The bird's-eye view could not follow North American cities into their eventual expansion, but remains a remarkable record of the first stages of American urban development.

CHAPTER SEVEN

City Maps and Plans

Gerald Danzer

Maps and plans of North American cities and towns document each stage and nearly every facet of urban history. Some of the information they convey resembles that of bird's-eye views, but maps and plans are in general more precise and specific. Each major city and most of the smaller cities and towns as well have maps and plans stretching back to their origins and marking the place's growth, decade by decade, up to the twentieth century. After the turn of the century, the documentation becomes almost overwhelming. A compendium selected from these maps, placed in chronological order and furnished with appropriate commentary, would seem to be a fundamental reference tool for understanding the development of any metropolitan area. But this type of atlas is rarely available for cities in the United States.

Two related reasons might be advanced to explain this neglect; first the nature of the maps themselves, and then the traditional mistrust of civic planning in the United States. A popular adage states it best: "The map is not the territory." Early maps of frontier communities, usually sponsored by developers, boosters, and real estate interests, were notorious for picturing dreams rather than realities. They were proposals for cities yet to be built rather than maps of places that actually existed. From the earliest period people looked at these maps as promotional pieces, subject to hyperbole and distortion. Yet, upon reflection, we must conclude that even these dream maps carry historical interest. Furthermore, every frontier community needed a plat to show the location of the lots that people were buying and selling. Land sales often preceded the development of streets and squares, so that the early maps were the only way people could visualize what the town might look like. Moreover, the surveyors' plats on which early maps were based (see Chapter 4) were essential documents, drawn with care and preserved as one of the fundamental records for each community.

As we have seen in Chapter 3, there were many city plans drawn in North America before the nineteenth century. During the eighteenth century, maps of colonial cities appeared with some degree of regularity for purposes of defence, administration, commerce, promotion, and celebration. Reproductions of many of these early maps are available in inexpensive format in the Historic Urban Plans series, described in the bibliography to this chapter. Boston was perhaps the best mapped city in colonial times, and this series has maps of that city dating from 1722, 1728, 1743, 1767, 1774, and 1777.

In the nineteenth century urban cartography was enriched by a variety of real estate maps, especially for cities in the West. These maps often circulated among eastern capitalists who bought and sold lots on the advice of agents. The investors did not need to make an arduous trip to a future Queen City of the West to benefit from its prosperity; they only needed a handy plat showing the various lots and subdivisions of the future metropolis. The tradition of these subdivision maps has, of course, continued down to our own times. Between 1850 and 1920 some cities were the subject of one or more elaborate real estate atlases that afforded systematic coverage of the metropolitan area. Many fire insurance atlases (see Chapter 9) also served as real estate references, while for smaller cities and towns the county atlases served the same purpose (Chapter 8). Few large metropolitan areas received a county atlas, but one conspicuous exception might profitably be consulted by every urban historian: the *Atlas of Cuyahoga County Ohio* by D. J. Lake (Philadelphia, 1874).

Street maps are another cartographic type that became increasingly common in nineteenth-century North American cities. They provided guides to help residents and strangers find their way about cities grown too large for the average person to know as familiar turf. Many of the early street maps appeared in directories or handbooks describing sites of interest and commercial services. Later, separately published street maps tended to emphasize streetcar routes and schedules, house numbering systems, and street names. These ordinary street maps can provide a remarkable chronological record of the development of the urban fabric. Many cities have annual editions of these ephemeral maps stretching back a hundred years or more. They differed from the general reference maps of the cities, tending to focus on the street pattern and to neglect other important urban features. They exaggerated the width of streets and were usually crudely drawn by cartographic standards. But an inexpensive format and widespread use made street maps guides for

the people, a fact that cultural historians will not wish to overlook. They may have played an unintended role, by providing a basic image of the city to several generations of urban dwellers. The fact that they were revised each year and were published over such a long period of time provides a useful record of the development of street patterns and transportation routes. But the image of the city that they conveyed to citizens, newcomers and strangers alike, also makes them valuable sources for understanding the nature of the urban experience in North America (see, for instance, Plate 7.6). It is a shame that few institutions have bothered to collect and preserve these maps in a systematic way. The highway maps of the twentieth century (Chapter 11) may be considered a continuation of this type of map, bringing the story up to the present day. Street maps and highway maps, then and now, made the expanding metropolitan area comprehensible and accessible to its citizens.

Another rich source of information on urban places is the series of topographic quadrangles published by the U.S. Geological Survey (see also Chapter 10). The wealth of cultural data found on these maps should not be overlooked by students of North American cities. Although not as detailed as the real estate and fire insurance maps, the quadrangles provide several unique advantages. First, they present an accurate image of the physical site on which the city rests. Secondly, many locations were surveyed at intervals from the 1880s to the present day, providing cartographic benchmarks from which changes may be observed and measured. Some cities have different base maps available, on microfilm, for the late 1890s, the 1930s, the 1950s, and the 1980s. Since all regions of the nation were surveyed by the U.S. Geological Survey, and the maps all employ the same standards and symbols, one can readily compare the physical shape and growth patterns of various metropolitan areas. Moreover, the inexpensive cost of the current maps, $3.60 each for both the 7.5- and the 15-minute quadrangles (see Chapter 10 for an explanation of these terms), makes them ideal maps for classroom use or field study. Although these maps have long been used in geographical studies, historians have been slow to see their advantages. Indeed some libraries made a practice of discarding the old editions of these maps as soon as new ones became available. Riley Moore Moffat has compiled a useful though incomplete *Map Index to Topographic Quadrangles of the United States, 1882–1940* (Santa Cruz, Calif.: 1986).

As North American communities developed cultural self-consciousness in the nineteenth century, retrospective historical maps came into vogue. These maps showed what the community looked like at a much earlier period. The map legends sometimes contained signed statements by early pioneers attesting to the accuracy of the information. In other cases there are affidavits indicating that the map is an "exact replica" of an early manuscript that has been lost. Students of the past can make two mistakes in dealing with these maps. The first is to take them too seriously, and the second is to dismiss them altogether. Instead, these maps should be regarded as any other secondary source: they are reconstructions of the past with the strengths and limitations of any such endeavor.

As cities developed in the nineteenth century a variety of special-purpose maps appeared to aid in the construction of the modern metropolis and to help people understand the dynamics of city life. The first large group of these thematic maps was produced by city engineers. They showed sewer systems, traffic patterns, flood hazards, freight facilities, parks, and a variety of public works. A second group portrayed legal districts and jurisdictions. Maps of wards, precincts, and census tracts, and police, fire, and school districts usually were official documents published by the agencies involved. A third group of special-purpose maps,

often compiled by social scientists and various study commissions, tried to record the spatial distribution of population characteristics, urban services, health conditions, urban problems, and the like. Most of these maps have an ad hoc character and they appear by chance rather than systematically. Each city has its own corpus of these documents and they are widely dispersed, but they often pay rich dividends to the patient researcher.

As an example of the special urban maps produced by the early social science movement, we could take the "Hull-House maps," generated by the Chicago settlement house of that name, and reproduced as Color Plates 3 and 4. Agnes Sinclair Holbrook, a resident of Hull-House who directed the project in 1893, wanted to portray the ethnic mix of a port-of-entry neighborhood. She used base maps of the neighborhood around Dearborn Station provided by a local fire insurance map publisher and then established fifteen categories, one for each nationality group in the area, and assigned each one a different color. On the basis of a house-to-house survey conducted over several months, she and her co-workers determined the number of people from each group living in each building. The numbers were converted to percentages for each structure on the map. Thus a tenement housing twenty people, ten considered to be Irish and ten Italian, would be shown half green and half blue. One cannot tell by the colors on the map how many people live at any particular address; one can only see the ethnic mix of its residents. One must also rely on the methods and categories of Holbrook and her colleagues. For instance, they distinguished between first- and second-generation immigrants, counting children sometimes as English-speaking while considering their parents as Germans.

Each city has a variety of general reference and special-purpose maps that are of interest to historians. Some were published separately as wall maps or as pocket guides, but most were issued in books, atlases, government documents, and reports of every description. In addition, manuscript maps often rest in governmental offices, institutional archives, and reference libraries. Finding and using these sources sometimes requires a great deal of ingenuity, but a simple suggestion might be in order: be sure to consult the people in charge of the map collections in local libraries, historical societies, and universities. The very nature of map publication, collection, and storage makes professional cartographic librarians indispensable even to the most advanced scholars.

Students also will not want to overlook obvious sources such as the large atlases produced in nineteenth-century North America by the leading map publishers: Colton, Cram, Mitchell, Rand McNally, and the like. Most nineteenth-century atlases included a variety of urban maps, sometimes over a hundred individual sheets. For instructional purposes these atlas maps are inexpensive, easy to reproduce, and available in a convenient size.

A final suggestion might be the most important: we should not think of maps and plans as distinct types of sources, standing alone and in isolation. Cartographic resources have the greatest value when used in conjunction with other documents. Each type of source informs the other, confirming its accuracy and reliability, suggesting its potential value, and encouraging the reader to ask new questions and to look for new information. To base one's argument on cartographic data alone, without recourse to other sources of evidence, is to seriously misunderstand the nature of maps as well as the historian's quest. The map, after all, is not the territory.

SOURCES FOR CITY MAPS AND PLANS

Banham, Reyner. *Los Angeles: The Architecture of Four Ecologies.* New York: 1971. A study of one city with ideas applicable to many others.

Carter, Harold. *An Introduction to Urban Historical Geography.* London: 1983. A general synthesis of recent scholarship arranged in a systematic textbook.

Clay, Grady. *Close-Up: How to Read the American City.* Chicago: 1980. This handbook for observing the contemporary city provides a vocabulary and a framework for looking at maps as well.

Historic Urban Plans. *Historic City Plans and Views.* Ithaca, N.Y.: 1985. A catalog of reproductions featuring historic maps of a variety of European and American cities. These reproductions are very useful tools for studying urban history, and the catalog can be obtained by writing to Historic Urban Plans, Box 276, Ithaca, New York 14850.

North American City Plans. (Map Collectors' Series number 20). A list of examples with several lines of commentary on each map. The volume is arranged by city and includes reproductions of many of the plans.

Papenfuse, Edward C., and Joseph M. Coale III. *The Hammond-Harwood House Atlas of Historical Maps of Maryland, 1608–1908.* Baltimore: 1982. Pages 91–132 present, in chronological sequence, the important maps of Annapolis and Baltimore.

Raitz, Karl B., and John Fraser Hart. *Cultural Geography on Topographic Maps.* New York: 1975. This handbook by two masters of the craft uses thirty topographic maps to illustrate a variety of geographic principles. Part VII of an accompanying *Study Guide* uses the maps to help students discover the geographic aspects of towns and cities.

Reps, John W. *Cities of the American West: A History of Frontier Urban Planning.* Princeton, N.J.: 1979. This is a huge volume of over 800 pages filled with maps and views of the western towns in their formative years. Another Reps work is *The Forgotten Frontier: Urban Planning in the American West before 1890.* Columbia, Mo.: 1981. It covers the same ground more generally.

_____. *The Making of Urban America: A History of City Planning in the United States.* Princeton, N.J.: 1965. With over 300 illustrations, most of them full-page photographs of plans, this classic volume remains the essential reference on the topic, placing any particular example in the context of the whole corpus of North American urban plans from the beginning up to the dawn of the twentieth century. The same material appears in condensed form in Reps, *Town Planning in Frontier America*. Princeton, N.J.: 1969 (reprinted in different format, Columbia: Mo.: 1980).

————. *Monumental Washington: The Planning and Development of the Capital Center*. Princeton, N.J.: 1967. The story of the capital's development raises central questions about the appearance and structure of the nation's cities.

Richason, Benjamin F. *Atlas of Cultural Features*. Northbrook, Ill.: 1972. This atlas uses aerial photographs and topographic maps to trace the imprint of human activity on the land. Part 6 deals with "Urban Impressions."

Scargill, D. J. *The Form of Cities*. London, 1979. A useful introduction to the forces influencing the form that cities actually took, as distinguished from the proposals set forth in the plans of the founders.

Schlereth, Thomas J. "Past Cityscapes: Uses of Cartography in Urban History." In Schlereth. *Artifacts and the American Past*. Nashville: 1980, pp. 66–86 and 257–63. This essay uses Chicago as a case study and suggests how various types of maps might be used by historians; the notes are rich in bibliographical suggestions.

Shelley, Michael H., ed. *Ward Maps of United States Cities: A Selective Checklist of pre-1900 Maps in the Library of Congress*. Washington, D.C.: 1975. This checklist includes information on thirty-five cities; it was compiled to aid genealogical research, but historians will find it useful.

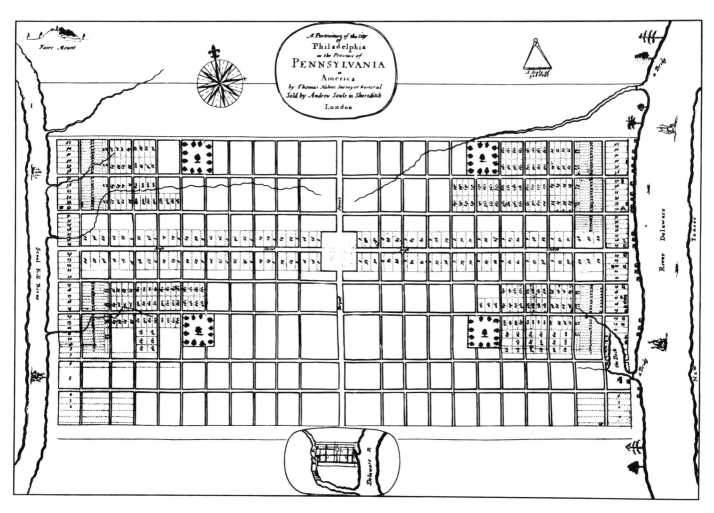

Plate 7.1. *A Portraiture of the City of Philadelphia,* by Thomas Holme (London, 1685).

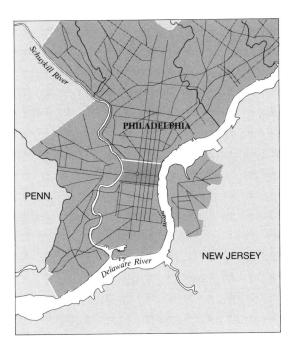

Plate 7.1a. Modern map to show area of Plate 7.1.

A Colonial City Plan: Philadelphia, 1683

William Penn was a dreamer with a difference: he had the resources to implement his dreams. Thus Philadelphia's early plan was not an idle speculation but became the blueprint for the major city serving Penn's territories in North America. Although the plan reflected the contemporary British approach used for town planning in Ireland, certain features looked back to an agrarian past while others looked forward to a utopian future. Sorting out these different directions is a challenging assignment, but holding them all together is the conviction that the shape of the settlement pattern would have social and political as well as economic implications.

One way to build a new society was therefore to build a new town. The site selected for the grid pattern of Philadelphia's streets was at the point where the Delaware and Schuylkill rivers were nearly parallel. The course of the rivers rather than the compass determined the orientation of the city. Individual blocks, lots, and streets were given different sizes according to their function. Major purchasers of land from Penn received, along with their rural holdings, lots in the city and a tract of "liberty land" in the area immediately adjacent to Philadelphia. Thus rural, urban, and suburban interests were joined in an effort to combine economic and political interests.

The plan had no special provision for places of public worship, an idea in keeping with Quaker theology and practice. Of the five public squares that break up the grid, four were to be used as parks, while the central one was reserved for public buildings. The central square served as the focal point for the plan, with Broad and High streets forming the major axes of the community. Each was one hundred feet wide, a huge size considering that the widest street in London at the time was only sixty feet. One historian has remarked that it took three centuries for Philadelphia to grow into its streets.

The scale of the plan says something about Penn's imperial stance. It would take over a century for the town to fill up all its lots. In effect, the founder thought he knew what would be good for the future. The use of large lots was one of these convictions, supported by a vision of an entire city of separate houses enclosed in gardens. The reality of the walking city dictated something different, and the large blocks and lots were subdivided after the American Revolution, adapting the original plan to popular taste. As a result the four recreational squares had streets added on all four sides, ruining the original idea of traffic control. The dividers at the top right also mark off one tenth of a mile, perhaps an early use of a metric concept.

Source: Anthony N. B. Garvin, "Proprietary Philadelphia as Artifact," in The Historian and the City, ed. Oscar Handlin and John Burchard (Cambridge, Mass.: 1963).

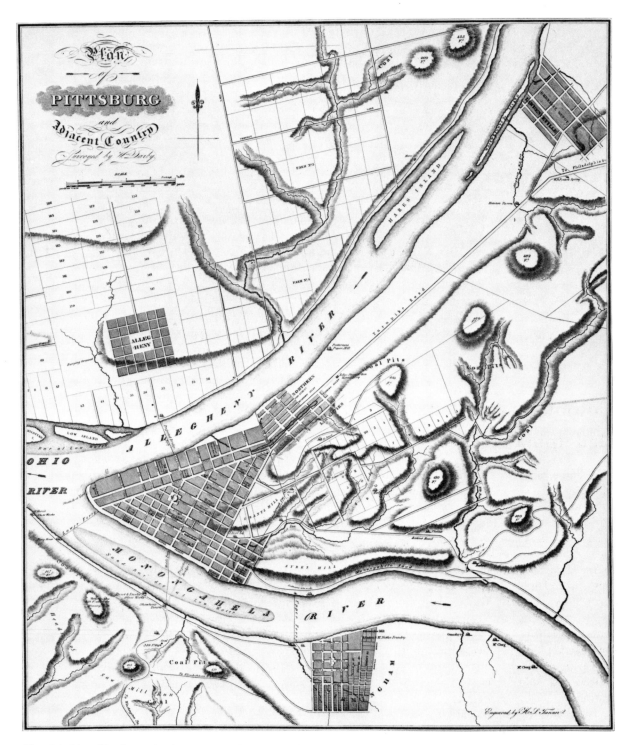

Plate 7.2. *Plan of Pittsburg and Adjacent Country,* by Wm. Darby (Philadelphia, 1815).

Plate 7.2a. Modern map showing the location of Pittsburgh.

A City Facing Westward: Pittsburgh, 1815

When the Treaty of Paris certified the United States of America as a new nation in 1783, almost all of its cities and larger towns hugged the Atlantic seaboard. To develop, the country had to extend routes for transportation and communication from the littoral to the interior. Cities had to be constructed at strategic sites on the way west and at crossroad locations on the frontier. The point where the Allegheny and Monongahela rivers joined to form the Ohio was certainly a critical place where the nation needed a city, as is clear from our accompanying map.

In 1784 several descendants of William Penn claimed ownership of the "Mannor of Pittsburgh" and prepared a plan for the development of the place. They retained the gridiron pattern of the old settlement that had grown up alongside Fort Duquesne and then Fort Pitt, and extended the design along the Monongahela. The blocks parallel the river's bank and the lots are laid out to face Water Street. The

proprietors then started a new grid design along two major streets laid out parallel to the Allegheny River. Liberty Street, which marked the break between the two subdivisions of the town, was 80 feet wide and was meant to be the major axis of the settlement. Note how the blocks on the north side of the city's triangle are oriented to Liberty and Penn Streets rather than to the river front. The hills deflected the street pattern as the Penn design was extended several decades later to a new area called Northern Liberties.

William Darby surveyed the *Plan of Pittsburgh and Adjacent Country* about 1815 and published it in Philadelphia in association with R. Patterson, a Pittsburgh merchant. It seems to combine the accurate measurements of the surveyor with the promotional qualities sought by local business interests. Note the coal pits south of both rivers, the extension of Penn Street, which became a turnpike to Philadelphia, the "projected" bridges across each river, and the glass factories, foundries, mills, and the wire works along the rivers. A steamboat yard is located at the point marking the source of the Ohio River.

The town of Allegheny, on the north bank of that river, reflects a different pattern of settlement. The large square at the center provides a distinct focal point for the community. It is surrounded by blocks of town lots, which yield to larger units of farmland after an intermediate space for future expansion of the town, or for the provision of public services. Note the burying ground located in this transition zone.

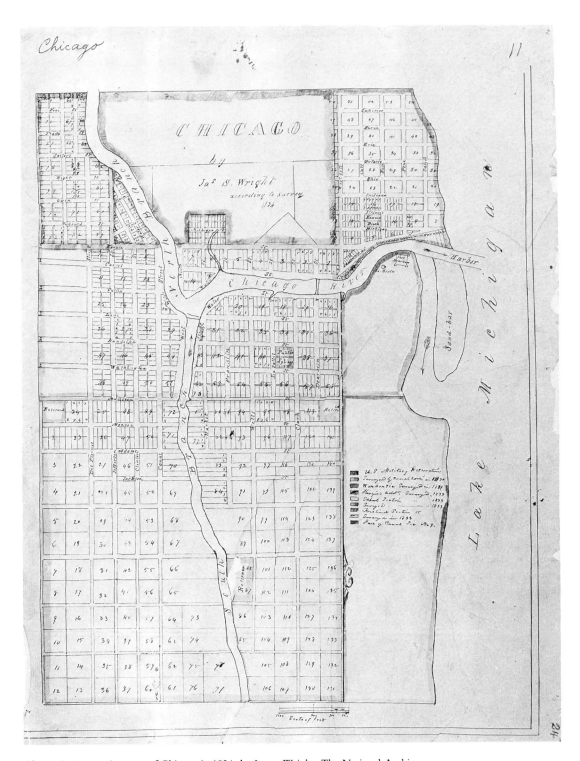

Plate 7.3. Manuscript map of Chicago in 1834, by James Wright. The National Archives.

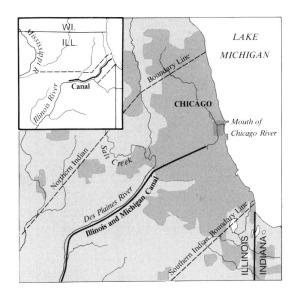

Plate 7.3a. Modern map to show area of Plate 7.3.

The Genesis of a Canal Town: Chicago in 1834

The genesis of Chicago roughly follows the pattern suggested by Frederick Jackson Turner in his frontier thesis of 1893. Native American people used the place and led the first European explorers, Father Jacques Marquette and Louis Joliet, to the site in 1673. Missionaries and fur traders followed with occasional visits, but the marshy ground at the southwestern tip of Lake Michigan did not become a permanent settlement until the period of the American Revolution. The trading post of Jean-Baptiste Point du Sable, erected in the late 1770s, was followed in 1803 by Fort Dearborn. The destruction of the latter in 1812 did not entirely wipe out the settlement that grew up around the military base. But even with the rebuilding of the fort, the laying out of a formal town had to wait until 1830, and the beginnings of the canal.

After the war of 1812 the Indians ceded a strip of land connecting the mouth of the Chicago River on Lake Michigan with the Illinois River in the Valley of the Mississippi; this strip was defined by the "Indian Boundary Lines." In 1827 Congress granted to the state of Illinois alternate sections of land in this highly strategic link to generate funds to dig a canal that would provide a water connection across the midcontinent divide. Three years later the canal commissioners hired James Thompson, a surveyor from southern Illinois, to lay out a town on their section of land next to Fort Dearborn. Chicago was born in section 9 of the grid established by the township and range system (see Plate 4.4). Its streets and lots largely, but not entirely, followed the surveyor's compass and chain. Thompson ran Carroll Street north of the main stem of the river slightly off the east-west axis to follow the left bank of the stream, an irregularity which persists to the present day. On the other hand, the river's channel has been continually reshaped in various places to fit into the grid pattern.

By the time that this map was drawn in 1834, additional parcels of land had been purchased and subdivided so that the town plan became a composite of different designs, some blocks running north-south in their subdivisions, and some east-west. Each part, however, was bounded by lines established in the original land survey. The one exception, Wonbonsia along the North Branch, did not last. North, South, and West Water Streets make the river a civic resource, providing public access to its entire course within the original town. Later subdividers were not so civic-minded, creating private lots on the river's banks.

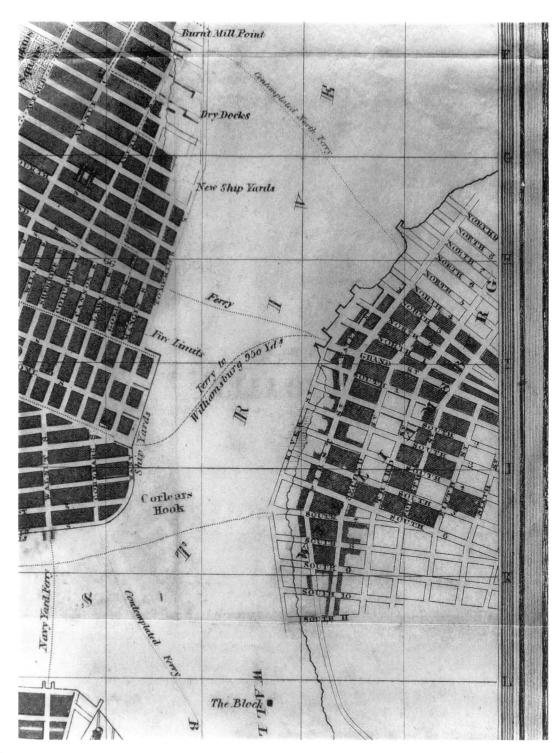

Plate 7.4. Detail from the "Map of the City of New-York drawn by D. H. Burr,"
in *A Picture of New York in 1846* (New York, 1846).

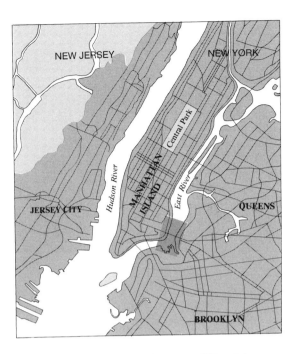

Plate 7.4a. Modern map to show area of Plate 7.4.

The Last Phase of a Walking City: New York and Brooklyn in 1846

The "walking city" characterizes the first state of North American city development. Most people walked to get around town. Since a comfortable walking distance was seldom over a mile, the physical extent of the walking city usually was limited to a circular pattern with a radius of only a mile or two from the city center. To conserve steps, people liked to work, shop, play, and worship all within a block or two of their residence. The land use pattern therefore called for mixing various types of buildings in every area of the city. Several functions were often combined in one structure. Lots were narrow, open space was in short supply, and individual buildings crowded together along streets and alleys. In the 1840s even New York, the largest city in the United States, could be characterized as a walking city, although it already had expanded to the point where ferries were a necessary part of its transportation system.

The detail shown here is from the "Map of the City of New-York drawn by D. H. Burr" for a guidebook called *A Picture of New York in 1846*. It is a plain black-and-white map, measuring ten by twelve inches, folded into a guidebook, and resembles hundreds of similar maps issued for North American towns and cities throughout the nineteenth century. The value of the map is vastly enhanced by the text of the 172-page guide that accompanies it. Thus we know that the Williamsburg Ferry ran every half hour from the feet of Grand and Houston Streets. The text, however, is silent on the "Contemplated North Ferry." Tompkins Square in the northwest corner of our detail gets an honest summary on page 85. It was one of the largest squares in the city, and found its major use as a parade ground. Yet it was "Not yet much frequented, as the trees are young, and the place not finished."

Instead of dwelling on the undeveloped square, the guidebook lists among the "places worth visiting" the new shipyard nearby. "Here may be found ships of the largest class, and steamers of every dimension . . . rendering it a series of infinite variety and interest. Extensive machine shops, for steam engines, will be found here" (p. 76). Williamsburg, the guide reports, was a "recently built town" that had expanded from a population of 5,094 in 1840 to 11,338 five years later. "Its chief buildings are a town hall and seven churches, together with handsome private dwellings" (p. 19). Thus the detail shown here catches the suburb in a boom period, the shaded portion apparently indicating the built-up area. Note how the plan of this suburb is oriented toward the shore line. The future right-of-way for River Street is projected out into the flats beyond the high-tide line (compare with Plate 3.9).

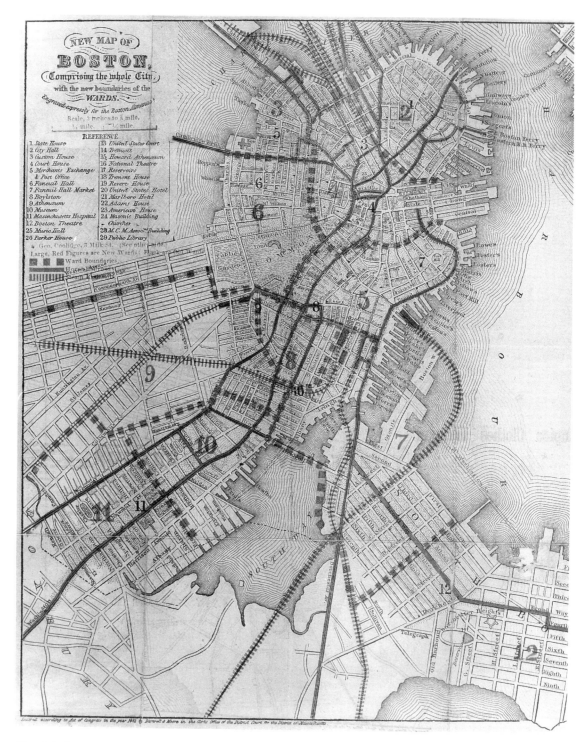

Plate 7.5. "A New Map of the City of Boston," from *The Boston Almanac for the Year 1866.*

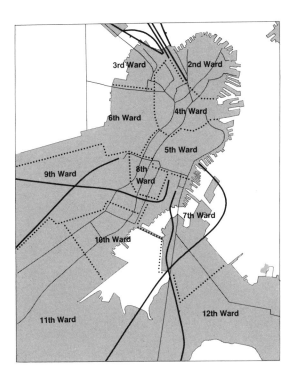

Plate 7.5a. Diagram
of Plate 7.5.

The Transition to a
Streetcar City:
Boston in 1866

This map is similar to the previous one, and accompanies *The Boston Almanac for the Year 1866.* We show the main portion of a less finely printed piece measuring eight by ten inches. One immediately senses some differences between the two maps. On the original of the Boston map, dotted red lines boldly delineate the new ward boundaries (dotted lines on Plate 7.5a) and seem to divide the city into distinct parts. The other overprinted lines trace the routes of "Horse Railroads" (thin lines) and "Steam Railroads" (thick lines).

The three-hundred-page book in which the map appears was part almanac, part citizens' handbook of government, part guidebook, and part advertising medium. It was meant primarily for local inhabitants, especially newcomers to the city, rather than visitors. It lacks the descriptive passages that make the New York map come to life, but contains abundant practical information, such as a complete schedule for Boston's various transportation lines.

Many of the engravings that accompany the volume show street scenes. Several of the most useful of these accompany a section called "Rambles in Boston," which was probably meant to orient immigrants to the community. The commentary highlights Boston's public works and points out "interesting historical subjects." The first ramble is a detailed explanation of how to use the horse railway lines. The map with this text provides a vivid introduction to the new streetcar city. The base map carries an 1861 copyright, the delay in publication perhaps reflecting the exigencies of wartime. By the end of the war, however, Boston had emerged as a streetcar city and demanded a new type of map that would clearly indicate its system of mass transportation.

New ward boundaries, new maps, and eventually a new kind of city emerged with the introduction of streetcars. Sam Bass Warner provided a classic description of this transformation in *Streetcar Suburbs: The Process of Growth in Boston, 1870–1900,* 2nd ed. (Cambridge, Mass.: 1978): "In 1850 Boston was something familiar to Western history and manageable by its traditions . . . . By 1900 it had become, along with many European and American cities, something entirely new, an industrial and suburban metropolis" (p. 3). Few documents illustrate the wrenching nature of that transformation as well as this humble map in a forgotten volume. It underscores one of the fundamental rules of historical cartography: "Even simple maps have stories to tell, but they will speak only when spoken to."

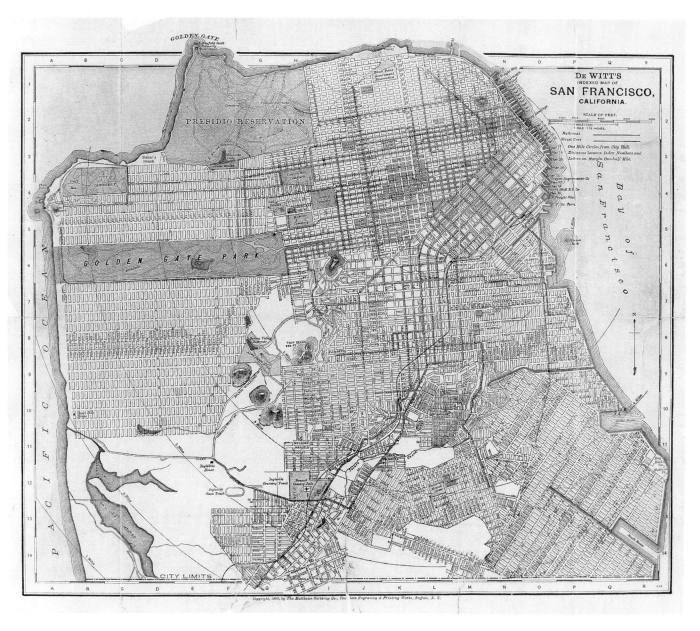

Plate 7.6. *Dewitt's Vest Pocket Map of San Francisco . . . Showing the Street Car Systems of the City, 1898.*

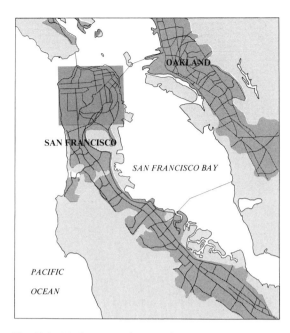

Plate 7.6a. Modern map showing the location of Plate 7.6.

The Full-Fledged Streetcar City: San Francisco in 1898

DeWitt's Vest Pocket Map of San Francisco . . . Showing the Street Car Systems of the City sold for fifteen cents in 1898. Its small size, ten by twelve inches, its descriptive title, and its basic structure all document an intended use by patrons of the streetcars. It shows the North American streetcar city imposing its structure even on the steep hills of San Francisco. Straight avenues proceed directly up the steep slopes and meet the cross streets at right angles.

The most pronounced lines on the map indicate the streetcar routes. The center of the system as well as the acknowledged heart of the city in 1898 was the city hall at the intersection of Market and Larkin Streets. Concentric circles mark off the mileage from the city hall. Only a small fraction of the streetcar lines—and the population as well—could be found beyond the three-mile radius. Although many streets

and blocks are indicated in the western and southern reaches of the city, everyone knew that these were only dreams until the streetcar lines reached out to incorporate the new subdivisions into the functioning entity of the city. Indeed much of the southeastern district was projected on lands then in the process of being reclaimed from the Bay.

Some streetcar lines extended out from the densely settled area within the three-mile radius to serve certain outlying areas: the Ingleside House and Race Track to the southwest, the large Golden Gate Park to the west, and Cliff House and City Cemetery to the northwest. The street patterns in the park and the cemetery as well as in the military reservation naturally followed the landscape. They were permitted, even encouraged, to break away from the gridiron pattern to provide a contrast to the regular city. Golden Gate Park broke up the regularity of the plan. Here one was expected to get off the streetcar and walk, or take a carriage ride through a different environment. The park was a place for reflection and relaxation, for meditation and cultural development, as well as for recreation.

At the time of this map San Francisco was nearly fifty years old. The town had developed from two focal points at the northeastern tip of the peninsula. Each of the original settlements provided a pattern for its own portion of the city. Market Street, following the line where the two patterns met, became the major axis of the city. Each new region seemed to develop a new pattern as it was added to the city. The far western subdivision, beyond Twin Peaks and the cluster of private cemeteries at the center of the urban area, represented the triumph of the streetcar style of urban development. Note how the streets and blocks follow the cardinal directions, stamped out in gridiron fashion to accommodate the streetcar tracks, and named according to the alphabet and numbers so that passengers would have no trouble finding their stops.

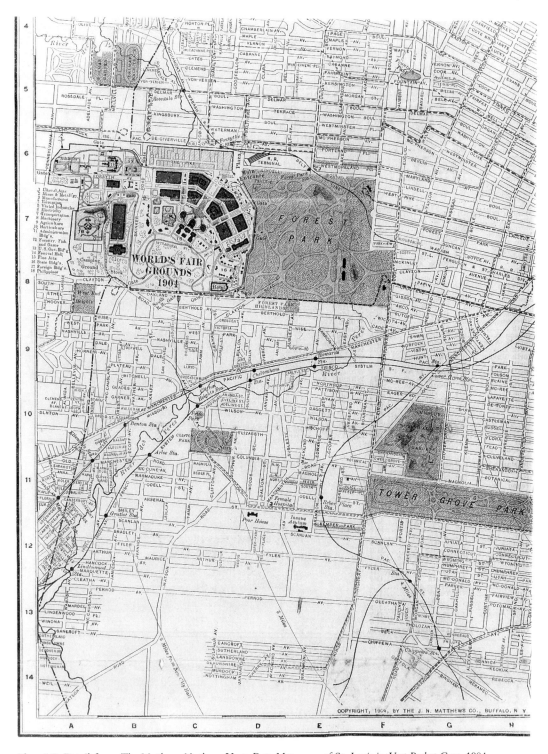

Plate 7.7. Detail from *The Matthews Northrup Up-to-Date Map . . . of St. Louis in Vest Pocket Case*, 1904.

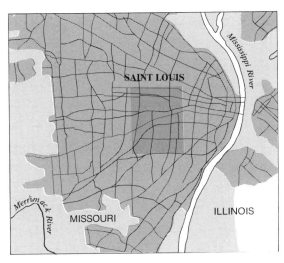

Plate 7.7a. Modern map to show the location of Plate 7.7.

The City Immediately prior to the Automobile: St. Louis in 1904

The Matthews Northrup Up-To-Date Map . . . of St. Louis in Vest Pocket Case sold for 10 cents in 1904, the date of its publication. Measuring a foot square, it was part of a series of similar maps available for eighteen major cities in addition to all of the states and territories. Although automobiles were no longer rare sights in St. Louis in 1904, the map still has the form of the streetcar city with little or no accommodation to the car.

In our detail some features of the railway city are evident. The World's Fair grounds at the edge of the city are tied to the metropolis by a variety of streetcar lines and by a major railroad terminus opposite the main gate. The grounds themselves are served by an intramural railway that weaves its way through the exhibition area.

South of the fairgrounds the commuter stations appear at regular intervals along the various railroads serving the city. Beyond the reach of commuter lines and streetcar tracks lie several large tracts of undivided land at the southwest margin of the city. At the edge of the built-up area are those institutions usually not welcome in more desirable locations: the hospital, the poor house, and the insane asylum.

As was often typical of streetcar city maps, concentric circles mark off the mileage from the center of the city. Forest Park is located about four miles from the city hall. Golden Gate Park in San Francisco, on the previous map, was located at roughly the same distance from the central business district. Both parks were sites for World's Fairs and both broke up the block plan of the city with their curved roads. Note how the railroad tracks crossed the northeast corner of the park to reach the fairgrounds. This was, one might object, an intrusion, an unfortunate example of the machine in the garden. But then we would be using the standards of the automobile age to judge the dynamics of an earlier city. Automobiles made it possible to spread out the metropolis even more, creating separate zones for every urban function.

The urban mentality that came along with the widespread use of private cars also tended to look with disfavor on the fixtures of mass transit systems. Tracks and terminals were often regarded as eyesores that impeded the flow of automobile traffic. Modernization demanded the replacement of streetcars with buses. Eventually the tracks in Forest Park were accompanied by a boulevard for automobile traffic. Still later the Daniel Boone Expressway cut across the southern edge of the park, and a giant traffic interchange extended eastward from the old hotel site.

The World's Fair grounds were divided up after the event, with the eastern part becoming part of the park. West of Skinker Road the land was given over to private residential and commercial development, though part of the land was used by Washington University and Concordia Lutheran Seminary.

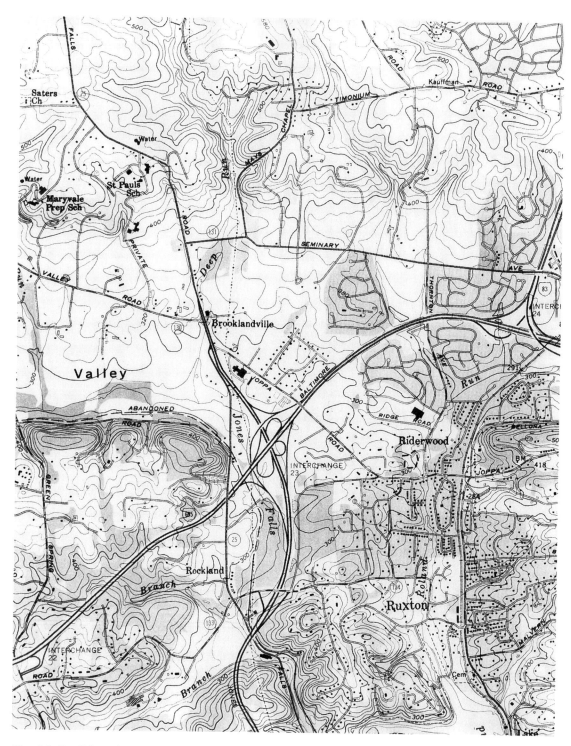

Plate 7.8. Detail from the USGS 7.5-minute quad of Cockeysville, Maryland (1974 edition).

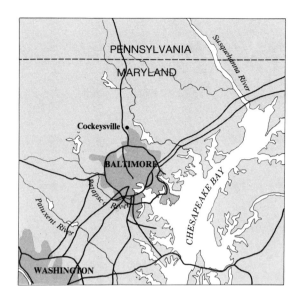

Plate 7.8a. Modern map to show the location of Plate 7.8.

The Triumph of the Automobile: An Area North of Baltimore in 1974

The transformation of the streetcar city into the automobile metropolitan region is graphically shown on the topographical maps issued by the U.S. Geological Survey (see Chapter 10). Ideally one would like to line up a sequence of these detailed maps, starting near the turn of the century and then proceeding to the late 1920s, the 1950s, and concluding with a contemporary map.

The Cockeysville, Maryland, quadrangle traces the development of an area north of Baltimore from a rural region in the early editions to the suburban sprawl featured in the detail shown here. The base for this 1974 map was developed in 1957. Additions and revisions based on aerial photographs were done in 1966 and again in 1974. Using all three versions one can trace the impact of the interstate highway system. The interchange shown here was only

partially built in 1966. The 1974 map indicates the new cultural features added to the map by printing them in purple. Thus the map indicates that Maryvale Prep School at the upper left was constructed between 1966 and 1974.

All of the subdivisions with the curved street patterns and the limited access are also shaded in purple. The squares indicate houses and one can see that these developments were just getting started in 1974 because very few houses had been completed when the aerial photographs were taken. The older settlement patterns feature straighter streets and provide more access points at the arterial roads.

The contour lines on the map indicate the nature and degree of slope of the land. Note how the curved roads in the older exurban development south and a bit west of the town name of Ruxton (at the lower right) closely follow the natural contours. The curved streets of the newer developments along the Baltimore Beltway to the north do not always follow the slope of the land. The area was level and the developer was free to press a preconceived plan on the landscape. Nowhere on this map is there any evidence of the grid system. In large measure this can be explained by the traditional pattern of land division in Maryland, in contrast to the township and range system used in the western states.

The original map indicates tree cover on the landscape by shading these areas green. Almost all of the steeper slopes in the area were wooded in 1974. At the bottom right a portion of Lake Roland, an impoundment of Roland Run, appears on this detail. It is surrounded by Robert E. Lee Park whose boundaries are indicated by a dashed line. The tracks of the Pennsylvania Railroad follow the eastern shore of the lake and extend northward to the expressway interchange. A branch line proceeding north and west from the other side of the lake is in the process of being abandoned, a telling comment, perhaps, on the triumph of the automobile city.

CHAPTER EIGHT

North American County Maps and Atlases

Michael P. Conzen

A special kind of map emerged in the United States and Canada during the nineteenth century to depict patterns of landholding and local geography at the scale of the county. Known commonly as *county maps* or *county atlases* and later also as *plat books,* this type of publication, almost always produced by private enterprise, was built on the principle of showing the ownership of all land parcels within the rural parts of a given county, set in the context of the area's main natural and cultural features such as rivers, hills, roads, railroads, towns, and administrative boundaries. The maps gained popularity because they cataloged the distribution and individual ownership of real property in an era of great population expansion and mobility, economic development, and social change.

This genre is significant because these large-scale maps usually represent a quantum leap over what had existed before in the detail and accuracy of local geographical knowledge, displayed in a graphic form that revealed the spatial pattern and coherence of rural neighborhoods and small-town settlements at a glance. Such information was important for merchants and farmers, owners of and traders in real estate, and anyone needing a ready reference guide of who lived where and what routes led where. Previous maps of American counties had been exceedingly rare, sparse in content, and limited in reliability. The new breed of map, which stressed the identification of individual rural residents or landowners, laid the foundation for newly accurate state maps around the time of the Civil War, and paved the way for more scientific government maps two decades later (see Chapter 10).

Begun as early as 1814, the idea of county landownership maps took three and a half decades to evolve into a reasonably standard form. Early experimental maps showed original land patent boundaries and grantees' names rather than current owners or residents; some showed both, some included topography and vegetative cover, others did not. The genre became fairly widespread after 1848 with the spread of lithography, and had its heyday between 1855 and 1925, when the character of maps and atlases became quite diverse in terms of visual design and supplementary content (see Plates 8.3, 8.6, and 8.11). The "golden age" can be said to have spanned the period from about 1850 to 1880, during which time the maps and atlases reached their peak of ornateness, detail, and supplementary content because they appealed to a mass market (see, for example, Plate 8.5). Thereafter, the genre redefined its clientele—giving up its sumptuousness to graphic publications of other sorts, in which expensive photography and color printing could be employed to the full—and addressed instead the needs of business, professional, and governmental interests, becoming more utilitarian in production and appearance. The Great Depression marked the end of many firms making this type of map, but the genre is continued by a few to this day, although in much stripped-down form.

There are about a dozen fundamental types of county ownership maps that can be fruitfully identified, depending on content, style, and market (a discussion of the technicalities behind this typology is included in Conzen, *Imago Mundi,* 1984, cited in the bibliography to this chapter). It is near paradoxical, however, that at the larger scale of concept and purpose a strong hint of formula publishing pervades the whole genre, while at the scale of individual mapmakers there are literally hundreds of minor variations so distinct as to give the impression of an infinity of ultimate products.

Before 1861 all landownership maps were of the large format, wall map type; after that the same basic information was more commonly packaged in an atlas, a township subunit of the county to each page, allowing additional pages for varied illustrative and textual material. Later changes in printing and marketing, however, produced several phases of renewed wall map activity, and while the atlases contained a wider variety of historical evidence, both types offered at a minimum the same basic mapping of detailed landownership patterns. There might be enormous differences in the appearance of particular county maps and atlases from one place to another and over time, reflecting various mapmaking skills, financial investment, and publishers' inclinations. At one extreme were the copperplate maps of counties

in eastern states during the 1840s, which exuded elegance and a highly professional mastery over the cartographic medium. At the other extreme one could find engaging examples of home-made maps, many of them from Iowa, with crudely conceived and executed house symbols and railroad lines with miniature railroad trains sporting gargantuan cowcatchers steaming right up the map—folk maps that signal the pervasive appeal of this kind of mapmaking across the country and throughout society. Nevertheless, many maps followed increasingly standardized formulas for content and presentation, and thus permit direct and valuable comparisons over space and time. It is this diversity in detailed layout, symbol, and manifest artistic skill in the finished maps that lends the genre an endless fascination for the map fancier and historian alike.

As is evident by now, these maps were made by numerous individuals from varied walks of life. Early on, professional surveyors tended to dominate the business—but surveyors trained or self-taught in an era when surveying still included and valued artistic creativity. There were many people who tried their hand at the genre. As maps and atlases included more supplemental illustrations, text, and statistics, the initiative passed to business organizers who could manage complicated publishing projects on a quasi–assembly-line basis and raise the requisite capital. By the 1880s comparatively few one-person outfits were left, and the largest firms issued tens and even scores of atlases a year. By the early twentieth century, however, cheap maps—unlovely to look at but functional, and loaded with advertisements—challenged the "full-service" atlases, and so the knockdown version came into widespread favor, again attracting many entrepreneurs at first because entry into the market had been cheapened, but fewer as time wore on and a trend toward bigness and consolidation again set in.

Throughout the entire history of the county landownership map the methods of production and marketing were essentially the same. The fundamental object was to make a product that by its nature would appeal only to a limited audience—the residents of the target county—so sales, by prior subscription, had to be maximized. The publisher had on the one hand to organize a work force to construct the map itself from the best existing sources, including new fieldwork and legal research in the deeds records, while on the other hand a sales force was needed to scour the county signing up potential buyers and selling them extra options, from their name appearing on the map in larger or bolder type than their neighbor's to lithographic views of their property, potted biographies, and portraits for inclusion in the work. Counties were chosen for "treatment," the maps were made, the lightning sales campaign conducted, the product produced and distributed, payments collected, and then it was

on to the next chosen county to repeat the process. In the early days, individual mapmakers carried out all these functions, and even when partnerships emerged between those with artistic talents and those with business acumen, rates of publication were slow. When it became a factory operation, map companies employed scores of people, and the largest moved their sales force around the profitable sections of the country like a miniature locust plague. Map peddlers, in common with all the other breeds of roaming hucksters and traveling salesmen, came in for their share of opprobrium as well as salutation.

Since the early nineteenth century it can be estimated that about 5,000 distinct maps and atlases have been published of numerous North American counties, sometimes at irregular intervals for the same counties. There are, however, several regions where county map-

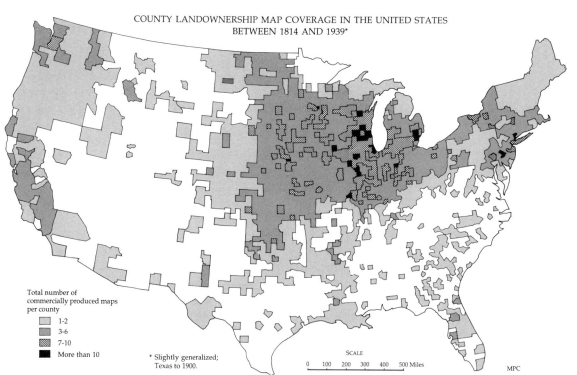

COUNTY LANDOWNERSHIP MAP COVERAGE IN THE UNITED STATES
BETWEEN 1814 AND 1939*

Total number of
commercially produced maps
per county

■ 1-2
■ 3-6
▨ 7-10
■ More than 10

* Slightly generalized;
Texas to 1900.

SCALE
0 100 200 300 400 500 Miles

MPC

Plate 8.1. Geographical occurrence of county landownership maps and atlases for the period 1814–1939.

ping was ill-developed or absent (Plate 8.1). County landownership mapping occurred in regions of rich, diversified yeoman farming. Early coverage from the 1830s to the 1860s is good for the Middle Atlantic states, New England, and Ohio. Counties in the Middle West, the Great Plains, Ontario, and the Maritimes were quite widely and frequently covered between 1850 and 1875, and those in the Middle West and Great Plains again, along with the Pacific Northwest and Canadian Prairies, between 1890 and 1925. In the profitable regions, this form of county mapping was repeated at as much as five- to ten-year intervals (by different mapmakers), creating a substantial cartographic record of the changing settlement landscape. By contrast, the American South, the Rocky Mountain regions, and Québec were sparsely covered, if at all, at the best of times.

Maps, particularly of this genre, project an aura of accuracy and authority that any self-respecting historian should suspect, not because it is hugely and frequently false (it is not), but because different components of these maps reflect different standards of care in compilation or states of knowledge at various times, and these limitations should be considered in evaluating these maps as sources of historical evidence. A preliminary review of county map accuracy has appeared elsewhere (Conzen, *Agricultural History,* 1984, cited in the bibliography to this chapter).

Suffice it to say that the many topographical details are reasonably reliable. Terrain and rivers may not be delineated with accuracy comparable to the USGS maps of the turn of the century (see Chapter 10), but the details are extremely useful if they are, as often they are, the only attempt at thorough rendering for the period. Residents' and owners' names are generally accurate enough to match up with censuses and rural directories. Missing cases are probably more likely than outright misassign-

ments, but no mapmaker was infallible. Western landownership naming, involving as it did cadastral boundaries marked on the map, tends to be more trustworthy than the residents' naming on eastern maps. Roads, houses, and other cultural features are usually quite reliable, though alignments and exact positions may be off somewhat (they were usually interpolated from other landmarks by eye from horseback or buggy in the field). In all, a certain amount of caution should be exercised until the characteristics of a given map or mapmaker have become familiar, maps by different makers should be judged independently. But these caveats aside, it remains true that these county maps offer much information that may fit into historical research in many different and useful ways, and that is retrievable from no other source and in no other way.

The research value of these maps lies chiefly in the extraordinary detail they present regarding the spatial organization of local agrarian society in North America from the 1850s to the 1920s. Since all rural landowners are identified on the map, along with considerable information concerning the topography, vegetation cover, transportation networks, industrial activities, urban places, churches, and schools in a locality for a particular date, these maps offer a full picture of the geographical basis of individual and corporate life in the districts concerned. For example, varying population densities can be deduced from place to place (Plates 8.3, 8.4, 8.5, and 8.6). Patterns of resource exploitation can be examined (Plates 8.3, 8.4, and 8.10). The urban process can be seen in internal city building patterns and fringe urbanization (Plates 8.9 and 8.11). Furthermore, the map evidence can frequently be combined with systematic historical material from a variety of other archival sources (Plate 8.6 and the works by Conzen and Norris cited in the bibliography to this chapter).

SOURCES OF NORTH AMERICAN COUNTY MAPS AND ATLASES

These maps and atlases are to be found in a wide range of national, regional, and local libraries and historical society collections, mostly in North America. There is yet no single comprehensive checklist of all known county landownership maps, but searches for particular items can be initiated through a number of specialized cartobibliographies, most of which pertain to the holdings of individual institutions. Since the existence of this type of map coverage varies so greatly from region to region and by period, it is desirable to begin by canvassing published national and regional bibliographies to determine whether a particular county is likely to have been well mapped or not, and where one needs to focus if extensive map coverage of this sort is needed. Maps and atlases are often listed in separate compilations because wall maps are usually stored in flat map cases and bound atlases on oversize shelves. This unfortunate bibliographic divorce must be overcome if map information for particular counties in either or both forms at various dates is not to be overlooked.

The exact number of county maps and atlases is not known, and research continues to uncover more and more cases of hitherto unheralded items. Many local libraries and collections have such maps for their respective areas without this being widely known elsewhere. For areas east of the Appalachian Mountains county maps and atlases are likely to depict house locations rather than contemporary cadastral information (that is, without delineation of individual property boundaries); in the areas of national land survey truly cadastral land ownership maps (showing property boundaries), with or without dwellings, should be the rule.

The largest single collection of county maps and atlases is found in the Geography and Map Division of the U.S. Library of Congress. Not only does this serve as the "national" collection, but it also has long received maps and atlases for copyright deposit, although this has not guaranteed copies of all material. The Library's pre-1900 maps are cataloged in Richard W. Stephenson, *Landownership Maps: A Checklist of Nineteenth Century United States Maps in the Library of Congress* (Washington, D.C.: 1967). A large set of microfiche reproductions of the maps listed in this source can be purchased, details from

the Library, and some major libraries in the United States and elsewhere have acquired the set. Post-1900 landownership maps in the Library's collection have not yet been cataloged, though the holdings are rich; enquiries about these holdings can be made, and photostat and photograph reproductions of county maps can be ordered from the Library. The Library of Congress collection of county maps is by no means complete, however, and more localized library searches can yield significant additional material for many areas.

County atlases are also cataloged in Clara E. LeGear, *United States Atlases: A List of National, State, County, City and Regional Atlases in the Library of Congress*, 2 vols. (Washington, D.C.: 1950 and 1953). Volume 2 is additionally useful in that it also indicates substantial holdings in other U.S. libraries at that time. Microfilm and photostat reproductions of county atlas material can also be arranged with the Library, and material from certain volumes may be photocopied at the Library. Again, neither volume captures all county atlases, since many further items have been reported since these lists were published.

A comprehensive collection of Canadian county maps and atlases is held by the National Map Collection of the Public Archives of Canada, and is accessible through two lists: Heather Maddick, *County Maps: Land ownership Maps of Canada in the 19th Century* (Ottawa: Public Archives of Canada, 1976), and Betty May, *County Atlases of Canada: A Descriptive Catalogue* (Ottawa: Public Archives of Canada, 1970). Various reproduction methods for material can be arranged. These Canadian guides appear to be virtually complete, for the smaller number of Canadian maps and atlases of this type have been easier to canvass.

Two other collections contain a surprisingly broad selection of U.S. and Canadian county maps and atlases. The British Library (formerly the British Museum) contains numerous North American county maps, many of them seemingly unique copies not in major U.S. collections. The relevant holdings of the British Library can be found in *Catalogue of Printed Maps and Charts and Plans*, 15 vols., with ten-year supplements (London: 1967 and continuing). Also, the American Geographical Society Col-

Michael P. Conzen

lection of the Golda Meir Library of the University of Wisconsin–Milwaukee holds a substantial number of early county maps and atlases of the U.S. and Canada, which can be identified from the American Geographical Society's *Research Catalog,* and through enquiry. Reproduction of material from both these libraries is possible.

Most official historical libraries in states and provinces where county mapmaking was important (e.g. the northeastern quadrant of the U.S., the Pacific coast, and Ontario) contain respectable collections of county maps and atlases, at least for counties in their jurisdictions. These collections may not always be cataloged, and special bibliographies are still rare. A superb new finding aid for county maps in one crucial region is Robert W. Karrow, Jr., *Checklist of Printed Maps of the Middle West to 1900,* 13 vols. (Boston: 1981); the additional *Index* volume (Chicago: 1984) separately identifies landownership maps throughout the checklist. Growing recognition of the research value of county maps in recent years has resulted in some handsome statewide cartobibliographies devoted to this genre of maps, such as:

Fox, Michael J. *Land Ownership Maps of Wisconsin.* Madison: 1978.

Miles, William. *Michigan Atlases and Plat Books: A Checklist 1872–1973.* Lansing, Mich.: 1975.

Treude, Mai. *Windows to the Past: A Bibliography of Minnesota County Atlases.* Minneapolis: 1980.

Metropolitan and local libraries, including university libraries, can sometimes yield useful collections of county maps and atlases of nearby areas. Individual county historical societies and public libraries in county seats should be checked if maps for a particular county are sought. Reproduction possibilities are more variable in these cases, but improving.

GENERAL WORKS ON COUNTY MAPS AND ATLASES

Conzen, Michael P. "The County Landownership Map in America: Its Commercial Development and Social Transformation, 1814–1939." *Imago Mundi* XXXVI (1984), pp. 9–31.

Ristow, Walter W. *American Maps and Mapmakers:*

Commercial Cartography in the Nineteenth Century. Detroit: 1985. Chapters 20–25 are particularly useful.

Stephenson, Richard W. "Introduction." In *Landownership Maps: A Checklist of Nineteenth Century United States County Maps in the Library of Congress.* Washington, D.C.: 1967.

Thrower, Norman J. "The County Atlas of the United States." *Surveying and Mapping* XXI (1961), pp. 365–73.

Walling, Henry F. "Topographic Surveys of States." *Van Nostrand's Engineering Magazine* XXXIV (1886), pp. 334–43.

DISCUSSIONS OF THE SCHOLARLY UTILITY OF COUNTY LANDOWNERSHIP MAPS

Conzen, Michael P. "Landownership Maps and County Atlases." *Agricultural History* LVIII (1984), pp. 118–22.

————. "Spatial Data from Nineteenth-Century Manuscript Censuses: A Technique for Rural Settlement and Land Use Analysis." *Professional Geographer* XXI (1969), pp. 337–43.

Sitwell, O. F. G. "County Maps of the 19th Century as Historical Documents: A New Use." *Canadian Cartographer* VII (1970), pp. 27–41.

EXAMPLES OF THE USE OF THESE MAPS IN HISTORICAL RESEARCH

Bowen, William A. *The Willamette Valley: Migration and Settlement on the Oregon Frontier.* Seattle: 1978.

Conzen, Michael P. *Frontier Farming in an Urban Shadow.* Madison: 1971.

————. "The Woodland Clearances." *The Geographical Magazine* LII (1980), pp. 483–91.

Norris, Darrell A. "Migration, Pioneer Settlement, and the Life Course: The First Families of an Ontario Township." In *Canadian Papers in Rural History,* vol. 4, ed. Donald H. Akenson. Ganonoque, Ont.: 1984.

Thrower, Norman J. *Original Survey and Land Division: A Comparative Study of the Form and Effect of Contrasting Cadastral Surveys.* Chicago: 1966.

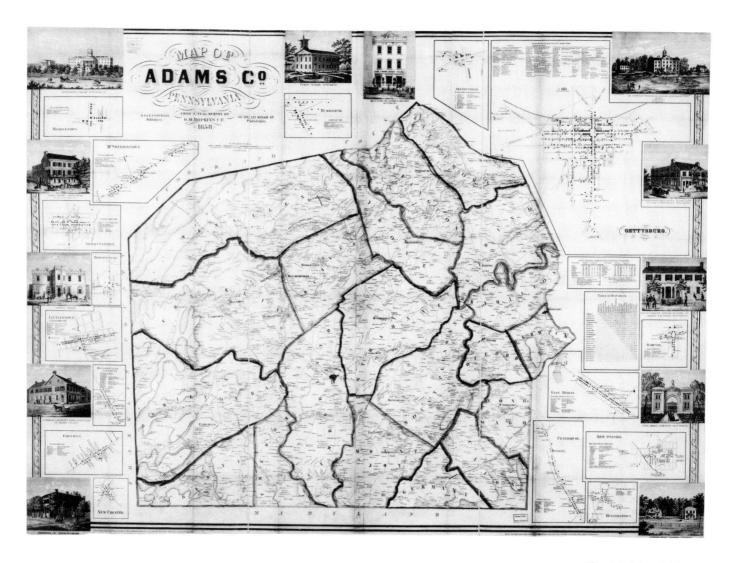

Plate 8.2. *Map of Adams County, Pennsylvania* (Philadelphia, 1858). The Library of Congress.

Michael P. Conzen

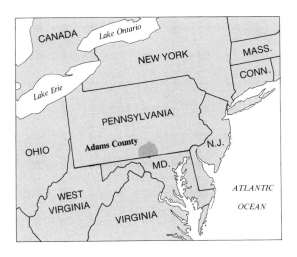

Plate 8.2a. Modern map to show location of Adams county.

Adams County, Pennsylvania, and Gettysburg in 1858

This is an excellent example of the classic wall map made of numerous counties in the eastern United States during the period 1845–1865. It represents above all a wholly new scale of information about the topography and settlement of localities, with new levels of detail and accuracy compared with what had gone before. Historically, this is the first map to show Adams County with all its country roads, farms, and isolated dwellings, hill ranges, river meanders, township boundaries, schools, churches, taverns, grist and saw mills, blacksmith shops, and other industrial premises as they were then—all adding up to a highly detailed portrait of how the local society was arranged within the territory of the country. Not as trigonometrically accurate as the USGS maps of the 1880s, this map nevertheless offered a much larger scale and far more types and quantities of physical and human detail than the "outline" county maps of previous decades, and it was the first time that virtually all private residences were individually identified on the map with the names of their inhabitants. Little wonder that Confederate generals, desperately short of maps at the outset of the Civil War, stopped at nothing to acquire copies of such publications.

The texture of this map is typical for areas of old metes and bounds land survey country (see Plate 3.4), highlighting the house locations of all resident rural families in relation to topography, roads, and towns. It is ideal for revealing the varying density of population from place to place, and why it varied. Given the convoluted property history of the eastern seaboard by the mid-nineteenth century, no attempt could be made to show actual property boundaries. Nevertheless, many people who figured prominently in national and local history can be traced to their parental homes and country estates on maps like this. The maps generally include large-scale inset plans of hamlets, villages, and towns with major buildings and residents identified.

This map nicely illustrates the site and built-up character of the county's seat, Gettysburg, five years before the Civil War battle. Lithographic views feature significant public, commercial, and private buildings, and offer generally reliable likenesses of the buildings. More tellingly, they hint at the pride and civic consciousness of county society. Distance tables and short lists of subscribers (usually businessmen and important local figures) complete the common formula. G. M. Hopkins, the map's maker, emerged from the tutelage of Robert Pearsall Smith, the noted Philadelphia lithographic map promoter, during the 1850s, and went on to become a specialist in the production of real estate atlases of large eastern cities (see Ristow, *American Maps and Mapmakers* [Detroit: 1955]).

Plate 8.3. Detail from the *Map of Norfolk County, Massachusetts,* by Henry Walling (Boston, 1858). The Library of Congress.

Michael P. Conzen

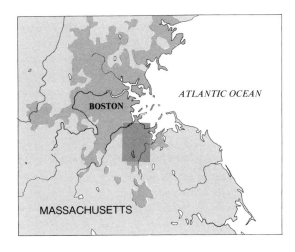

Plate 8.3a. Modern map to show location of Plate 8.3.

Rural Homes in Eastern Massachusetts

Walling's map of Norfolk County, Massachusetts (1858) shows in close-up the kinds of topographic detail the early county maps offered. This portion of the map focuses on the village of Quincy, nestled between Massachusetts Bay to the northeast and the Blue Hills to the immediate southwest, forming the southern rim of the Boston Basin. The terrain is represented by hachures depicting hill slopes, and all water courses are shown with characteristic meander patterns, flowing sometimes through marshy areas close to the coast. This terrain is drawn with only schematic accuracy, and should be checked against modern topographical maps, of the kind described in Chapter 10.

The road pattern is probably close to complete, since these maps were designed to assist travellers reach specific destinations. Typically, all residences are shown by small dots, most of them at roadside, and in the country areas the names of practically all residents are written on the map next to their homes. Village densities clearly precluded this, but then many villages are given detailed inset plans on the margins of such county maps. Occasional cemeteries, mills, and rural shops are also identified. The clustering of names helps dramatize the empty areas of the region, owing to its hilly state, though many families eked out a living on the lower slopes of the Blue Hills.

Quincy was throughout the nineteenth century a nationally important center of the quarrying industry, and its locational pattern is well displayed on this map. Granite quarries high in the hills (just above and below the name "West Quincy") drew laborers, many of whom were Irish, whose presence accounts no doubt for the Catholic Cemetery in the neighborhood. While the colonial village center ("Quincy Village") hugs the coast and caused the Old Colony and Fall River Rail Road to make a bee-line for it out of Boston before turning due south, there is Railway Village to the west to explain. This settlement formed where an established country lane intersected the gravity tramway which ran from the Railway Quarry north-northwest to a bend in the Neponset River. The first railway in North America (1826), this tramway helped develop a vigorous granite trade, which gained prime government contracts to furnish the grey stone for most of the major customs houses in the ports of the United States. And in North Quincy, Mrs. Billings, Mrs. Adams and others were neighbors, with the John Quincy family, to a cluster of slate quarries.

Henry Francis Walling, a major American topographical mapmaker, is aptly discussed in Walter W. Ristow's *American Maps and Mapmakers,* Chapter 20, pp. 327–38.

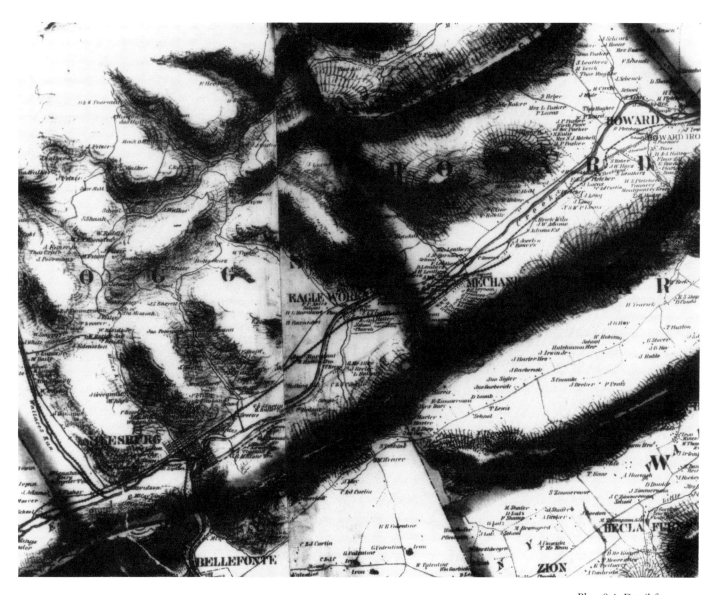

Plate 8.4. Detail from *Topographical Map of Centre County, Pennsylvania,* by Henry Walling (n.p., 1861). The Library of Congress.

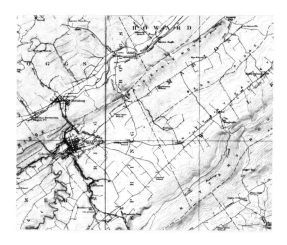

Plate 8.4a. Details from USGS 15-minute quads of Centre Hall (1929 edition) and Bellefonte (1909 edition), Pennsylvania, to show area of Plate 8.4.

Plate 8.4b. Modern map to show location of Centre County.

Vanished Iron-Working Landscapes Near Bellefonte, Pennsylvania, 1861

When Henry Walling published a *Topographical Map of Centre County, Pennsylvania* (1861) he was principally concerned to convey as accurate a picture as possible of what the region contained. The Allegheny Mountains run dramatically through this country from northeast to southwest, and Walling experimented with a form of hill shading to accentuate the impression of physical relief. This comes off rather well, but relative distances were hard to capture. Sand Ridge (lower right, north of Zion) appears as prominent as Bald Eagle Mountain, whereas it is little more than half the latter's local height. The difficulty of movement through the region is well illustrated by the angularity of the transport network of roads and railroads, confined to the valley floors and a few strategic gaps in the ridges. Hence the significance of the location of Bellefonte and Milesburg as classic "gap" towns.

Even more impressive is what the map reveals about the area's former livelihood. Bald Eagle Creek, parallel with the ridge on its northern flank, was crowded with iron-working settlements and related industry. "Howard Iron Works" boasted in 1861 a furnace and rolling mill, a tannery, and flour-, saw-, and plaster-mills. Eagle Works had a forge and a furnace, with rolling mills nearby. Elsewhere, iron works between Bellefonte and Milesburg and at Hecla Furnace suggest the importance of metalworking as well as farming at that time. Only seventy years later, all signs of this industrial activity have vanished from the USGS map. Mechanicsville became Mount Eagle, Eagle Works became Curtin, and Hecla Furnace became Mingoville. Many a mountain creek lost its mines and mills, leaving the area to farmers in the valley bottoms and second-growth trees on the heights.

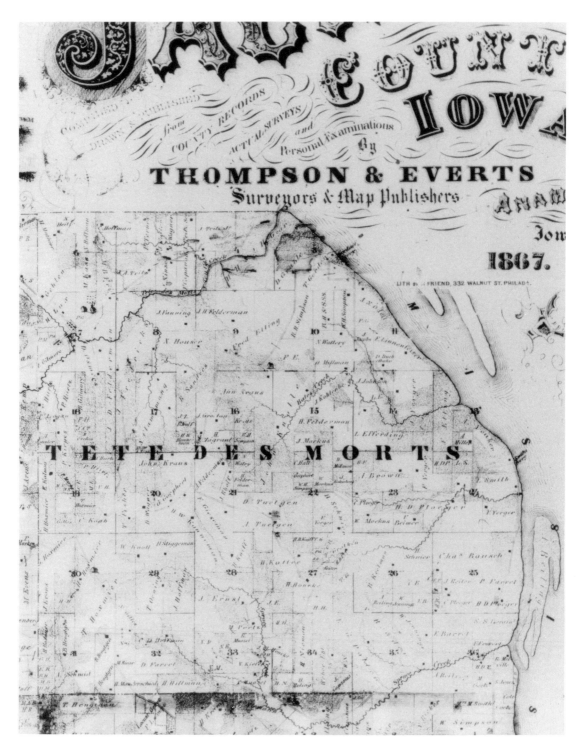

Plate 8.5. Detail from *Map of Jackson County, Iowa,* by Thompson and Everts (Anamosa, Iowa, 1867). The Library of Congress.

Plate 8.5a. Modern map to show location of Plate 8.5.

Farm Patterns South of Dubuque, Iowa

West of the Appalachians cadastral maps like this showing property owners (as distinct from residents) were common. The names appearing on such maps include nonresident owners and exclude tenants, and betray their mode of preparation, based on research in the county registry of deeds rather than on door-to-door investigation.

One can derive a good deal of information about settlement history by analyzing the nomenclature. In the case of the Jackson County map (1867), the names on its face reflect three phases of occupance. Tete des Morts (*sic*) refers to skulls the French travelers saw in this vicinity in the eighteenth century. The survival of this name has influenced many mistakenly to regard the main wave of farm settlement in the nineteenth century as French. Second, a scattering of Anglo-American property holders' names can be noted along the banks of the Mississippi, suggesting early nineteenth-century settlement, while third, the majority of farmers' names are of Luxembourg or Eifel-German origin.

The map symbolizes the era of woodland clearing and farm making, catching the area in midprocess, with cleared land most in evidence on the ridges between the streams and in the similarly flat valley bottoms. The pattern of farm boundaries in relation to woodland cover vividly portrays the variable progress made by farmers with different-sized holdings in clearing fields, and the farm shapes suggest attempts to gain land with a variety of slopes, soils, and drainage. The road network beautifully defines the linkages between upland and lowland rural neighborhoods that will have shaped the pioneers' daily lives and social patterns, as did the school locations for the children.

The lack of a river port in the locality indicates the farmers' land orientation to market towns like Dubuque to the north. This is somewhat surprising, as the ferry near the north end of the township offered connection with Galena, Illinois, directly across the Mississippi River. The ferry was not to last long. The only village in evidence is in the extreme northwestern corner of the township in section 7, with a church nearby. Platted in 1861 in a pattern reminiscent of old-world nucleated villages, the village's name, St. Donatus, is oddly missing from the map, suggesting that six years had not yet brought confidence in its perpetuation.

North American County Maps and Atlases   *199*

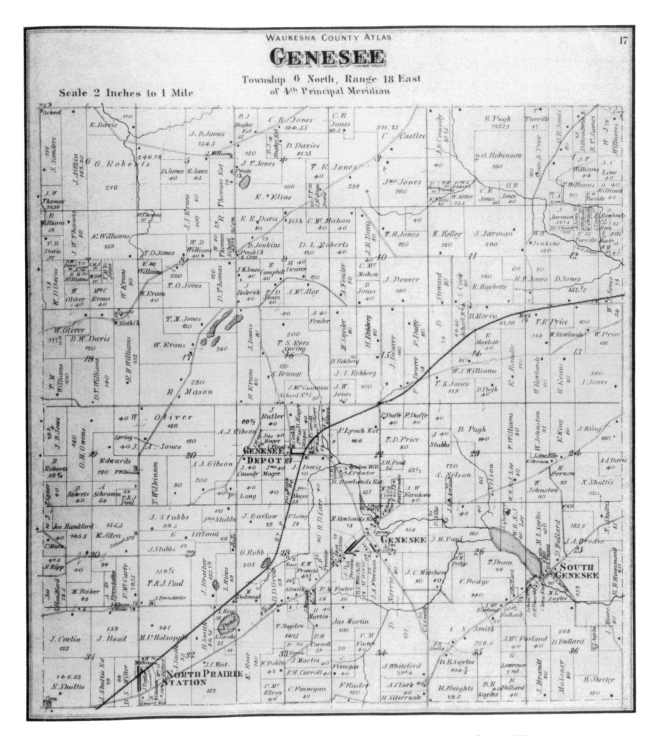

Plate 8.6. Genesee Township, from Harrison and Warner's *Atlas of Waukesha County, Wisconsin* (Madison, 1873).

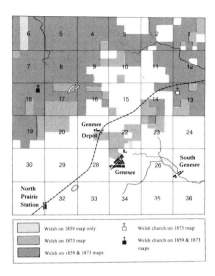

Plate 8.6a. Changes in Welsh landown-ership, 1859–73. Researched by Anne K. Knowles.

A Growing Ethnic Community in Rural Wisconsin, 1859–1873

If the evidence of names on a plat map can re-veal an ethnic settlement, a series of plat maps can set it in motion and—with a little help—bring it to life. Ethnic groups commonly bought and sold land among themselves, a practice the maps unveil as one parcel of land carried a sequence of, for example, distinctly German or Norwegian names over many dec-ades. Even by simply comparing names and farm sizes on plat maps over time, one sees whose wealth literally spread across a commu-nity, and whose withered or moved elsewhere.

The glimpse of names is tantalizing, but how to be sure of people's ethnicity, and how to trace their persistence or departures over time? The maps usually give first initials only, and ambiguity surrounds some common names in regard to ethnicity. One ready answer is the decennial U.S. census, which began to list the birth region of every U.S. resident in 1850,

along with much other useful information. Leafing through the manuscript schedules of the federal census for Genesee Township, lo-cated in the hilly Kettle Moraine region of southeastern Wisconsin, one finds strings of families from Wales. The accompanying ana-lytical map shows the result of combining cen-sus information from 1860 and 1870 with Gene-see plat maps for 1859 and 1873. Certain now of individuals' ethnicity, calculation reveals that the amount of township land in Welsh hands had increased from 33 to 43 percent by 1873. Welsh families made up 33 percent of the town-ship's population in 1870, up from 29 percent in 1860—a 4 percent shift, yet the number of Welsh landowners jumped by 16 percent.

The swelling Welsh community was not only filling in its borders, acquiring land from less tenacious or less interested English, Irish, and American farmers; its membership was also beginning to change, with longer-set-tled families emerging as significant land-holders. One early arrival, Richard Mason, moved from 40 acres in section 1 (1859) to 280 acres in section 17 (1873). A fourth Welsh church was established to service congregants in the eastern portions of the township.

At the same time, newcomers and the second generation were finding land increas-ingly scarce and expensive, resulting in more small farms. Yet the social advantages of living among native Welsh speakers, attending all-Welsh churches, and seeing their children marry "in the race" kept the Welsh strong in Genesee for several generations. As late as the 1920s, a young woman wrote in distress to one of her fellow Welsh friends that the Germans were "invading" the township—for the first time, a German family had bought an adjoining farm. The village of Wales was incorporated in the heart of the Welsh colony in 1882. Although it is now flooded with developments for Mil-waukee commuters, a few of the old families persist, as one would find in former ethnic farming communities across the Middle West.

Plate 8.7. Columbus, Ohio, from *Caldwell's Atlas of Franklin County, Ohio* (Columbus, 1872).

The Urban Pattern of Columbus, Ohio

County maps and particularly atlases can be excellent sources for large scale urban plans, in addition to the separate city maps that abound. Sometimes, if coverage of specific places is desired for particular dates, county landownership atlases may be the only source available. Often, such maps show only the basic street and block system without showing buildings. This grows out of the tradition that county maps and atlases were originated to show rural patterns in detail, and that urban places could be presented in summary fashion. However, in many instances, these sources yield superbly detailed information about the structure of the built environment.

In the case of central Columbus, Ohio, in 1872, the map shows not only the basic street grid, but all buildings too, with liberal annotation of the names and uses of countless public structures, the names of real estate subdivisions, and occasional ownership of grand houses. Not only is the Institution for the Deaf and Dumb identified, but we learn that the then Governor lived next door, at the head of State Street—the Pennsylvania Avenue of Columbus, as it leads westward to the edge of the State Capitol grounds at the left margin of the map. The siting of the Governor's mansion next to the institution makes sense in the light of state ownership of the city block on which they stand, but the asymmetry of the parcels suggest that a thin sliver of land along Oak Street was sold off at some time for private development. Furthermore, Governor Swayne had to endure overlooking a commercial skating rink in the next block, in a building that by its bulk and street building line must have dominated the view.

Elsewhere, the map gives abundant evidence of the ambiguities of American urban development history. Mr. Alfred Kelley owned a large portion of the city block fronting on Broad Street, in the center of which his mansion proudly faced the world, flanked by two other, less grand edifices. But what of the surrounding neighborhood? Modest houses on quite narrow lots appear the norm in most directions. What these juxtapositions say about intended residential status of districts and their subsequent history can only be hinted at here.

Near the lower left corner is a large market house, faced across the street by obvious business buildings built in a row occupying the tiny lots. That this area of town was full of churches, schools, a timber yard, fire station, and the like, besides residences, indicates the heterogeneous land use and mixed social patterns that existed in medium-sized American cities of the period. Lastly, as an indication of what the micro-morphology of the city can reveal, there are the numerous rear alleys of the district—a garbage-pickup paradise and a boon for automobiles when they eventually arrived. What accounts for these ubiquitous alleys in central Columbus? Alleys are far from ubiquitous in American cities. And why was there no horsecar service on any street save Long Street? Maps like this can spell and quell a host of questions about the way urban life worked in particular places.

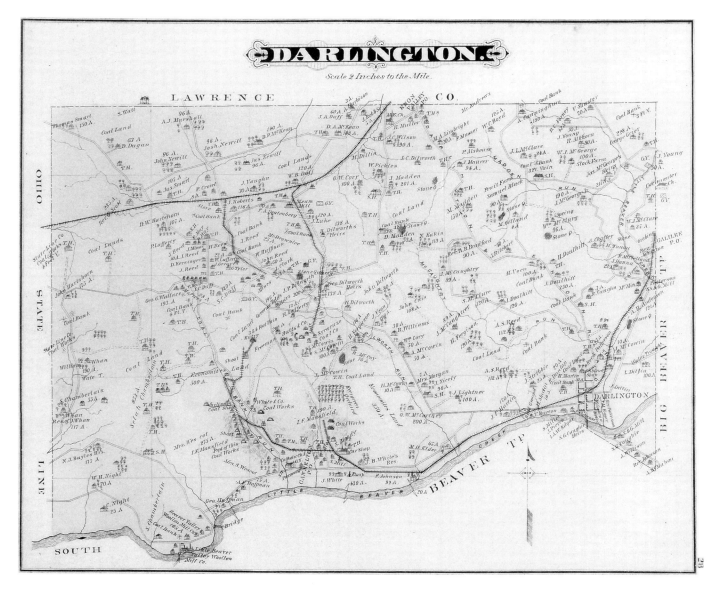

Plate 8.8. Darlington Township, from *Caldwell's Illustrated Historical Centennial Atlas of Beaver County, Pennsylvania* (Condit, Ohio, 1876). The Library of Congress.

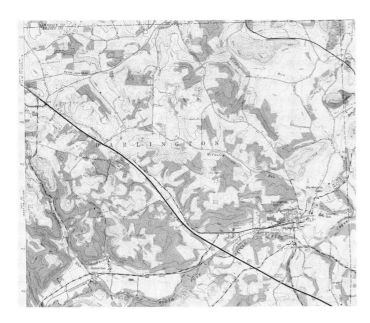

Plate 8.8a. Details from USGS 7.5-minute quads of New Galilee, Pennsylvania (1969 edition), to show most of area of Plate 8.8 in modern times.

Coalfield Landscape in Beaver County, Pennsylvania

The environment of western Pennsylvania offered only dispersed opportunities for farming in the nineteenth century, but the hills contained bituminous coal and oil in abundance. This map of Darlington Township in Beaver County shows the influence of this heady bonanza on the settlement pattern as it had evolved by 1876. Farms and other buildings are shown by pictogram house symbols, occupied by owners where names appear adjacent, and serving otherwise as tenant houses (T. H.). Although property boundaries are not shown, acreages of holdings are given, as well as descriptions of coal outcroppings (with, e.g., a 4-foot vein or such). Given the steep topography of the region, this map depicts the dense human occupance of the narrow valleys, with railroads straining to reach the isolated coalmines, and frequent school houses (S.H.) and occasional orchards, stone quarries, and mills—run in this district, obviously enough, by steam (S.M.).

A pattern of small landholders competing with large coal companies in these tight valleys at this time is also apparent, as is the common occurrence of graveyards (G.Y.). Out of this volatile economy came great profits that built substantial homes in the area, as the lithographic perspective views that accompany this map in the county atlas attest. The busy exploitation of resources is well mirrored in the ebullient, almost "folk" quality of this map's cartographic style. Such sources can contribute valuably in reconstructing the economic and social history of resource and industrial regions like this.

A century later, the mining towns of Cannelton and Darlington are still there (Plate 8.8a), though they have grown hardly at all. The dispersed miners' shanties are still strewn about the landscape, but the sites have all been moved about. Strip mining has removed practically all the remaining coal in recent years, and the wooded side valleys seem to lick their open-pit wounds and nurse a bypassed rural population as improved roads course through the area, and hint of a transformed economy.

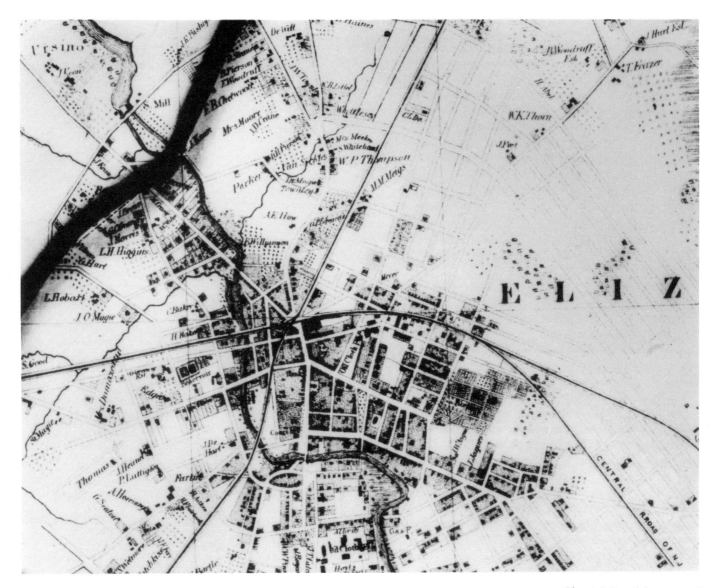

Plate 8.9. Detail from *Map of Union County, New Jersey,* by Meyer and Witzel (n.p., 1862).

Plate 8.9a. Modern map to show location of Elizabeth.

Urbanizing Village, Elizabeth, New Jersey

Some county maps were executed with such skill as to capture at one moment in time evidence of the broad changes in economy and society associated with urbanization. Elizabeth was a traditional village serving an agricultural hinterland up to the mid-nineteenth century. Here one can see an informal village street plan, filled with houselots, buildings, gardens, and the like, set in a sea of orchard farms (some with old Dutch family names) strung out along the country roads.

By 1861 the area had been rudely traversed by the "iron horse," now sporting stations sited as much to receive (and indeed encourage) suburban expansion as to serve the orchardists. The town's street system is in process of expansion in all directions, and particularly into the marshes to the east. The town was acquiring a lively but still separate port district on the Newark River, with a smattering of factories (glass, varnish, ropewalk) within the optimistic new subdivisions of the Elizabeth Port Land Improvement Company. The great marsh to the northeast would remain free until colonized by the modern Newark International Airport. Already the proceeds of industrialization are evident in the fancy country estates sprouting on Elizabeth's periphery, such as Ursino to the northwest. Unsurprisingly, the poorhouse was relegated to the fringe of the marsh east of North Park. The entire area has since been urbanized, and vast tank farms now stand where the orchards east of Wheatsheaf once flourished.

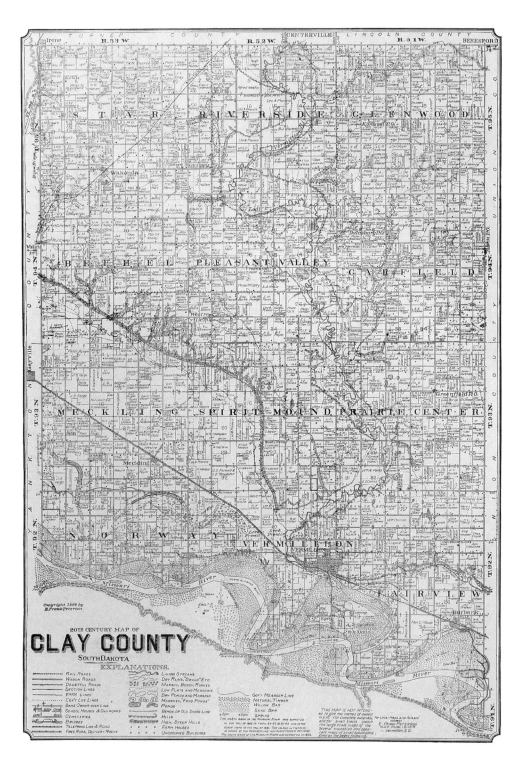

Plate 8.10. Detail from
*20th-Century Map of Clay
County, South Dakota,* by E.
Frank Peterson (Vermillion,
South Dakota, 1900).

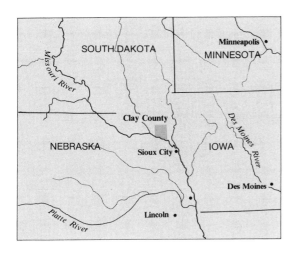

Plate 8.10a. Modern map to show location of Clay County.

Agricultural Lands in South Dakota

By the end of the nineteenth century landownership maps were fast losing their aesthetic qualities. This example, a map of Clay County, South Dakota, in 1900, shows the trend towards the functional, but also illustrates the way in which a single mapmaker could create an individual style, even under these circumstances. Compared with other county atlas makers of the time, E. Frank Peterson, headquartered in Vermillion, South Dakota, and committed to the complete blanketing of his state with county atlases, evolved a busy style that filled up the map with details others did not care to include. Although not the sole cartographer to show such information, he included the routes of telephone lines through the countryside, the pattern of rural free (postal) delivery, as well as highly detailed physiographical data on hydrology as well as terrain and vegetation, with particular attention to springs. In addition, his detailed township maps included measurements of all section lines for the calculation of accurate acreages in places where the original land survey may not have marked off perfect forty-acre parcels.

The attention to precise details concerning physical features shows particularly well in the case of the Missouri River and its historical meander movements. Problems quickly arose when rivers shifted away from the margins they had when government surveyors first went through the area, and inconvenienced landowners would have to seek costly legal adjustment of records and plats in order to avoid being wiped out by nature the slow way. The opportunity to compare a careful county map by Peterson with the original government maps and with later resurveys allows considerable physical change to be related to the social and economic history of the county.

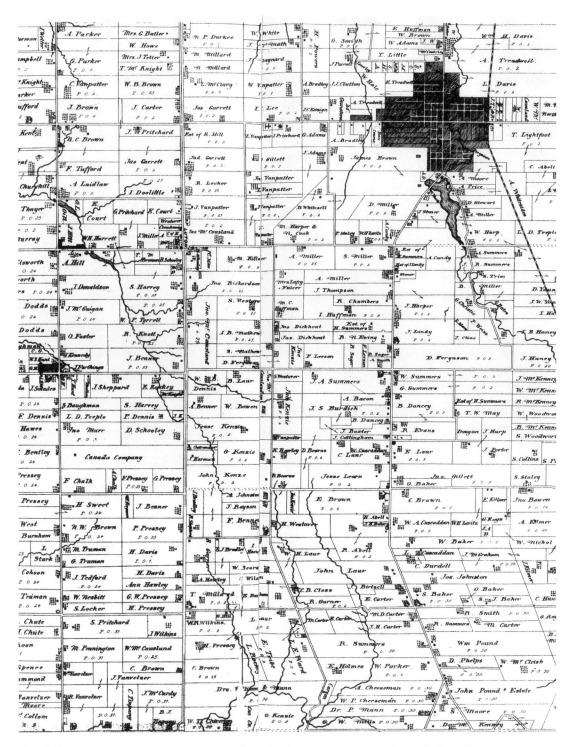

Plate 8.11. Detail from the *Illustrated Historical Atlas of the County of Elgin, Ontario,* by H. R. Page (Toronto, 1877).

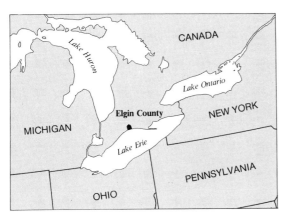

Plate 8.11a. Modern map to show location of Elgin County.

Rural Patterns in Ontario

A segment of the Elgin County, Ontario, map of 1877 by H. R. Page of Toronto reminds us that Canadian land division customs are somewhat different from those in the United States, and have produced in eastern Canada a different land geometry. Mapped at a similar stage of colonization as the districts presented in Plates 8.5 and 8.7—that is, about twenty to thirty years after initial main-wave settlement—this portion of Malahide Township displays resolutely rectangular farms neatly arrayed along well-behaved country roads. How well does the deference to government-imposed order in southwestern Ontario stand in contrast to the higgledy-piggledy, individualistic mosaic of American farms in eastern Iowa!

Further examination of the map reveals other distinctions. Areas of woodland are not shown, but orchards are. At first glance it would seem that everyone was blessed by the cartographer with an orchard, like a party favor, and that confidence might have to be withheld concerning the incidence and placement of these plots. Their size is symbolic, but their existence can be checked in the census, and orchards were not only a great craze in the 1870s but also climatically favored in this part of Canada. Mr. Page, moreover, had apprenticed with Captain Andreas, the celebrated county atlas-maker of Chicago, who had introduced orchards to atlases to signal that the country—wherever it was—was becoming civilized.

The pattern of rural services emerges very clearly from a map of this character. For example, the town of Aylmer clearly dominates the district, but there is at least one hamlet on the map, Luton, to the southeast, that offers some central place services. While all farmers undoubtedly went to Aylmer to market their farm produce, since the town had the railroad, journeys to pick up one's mail would be more sensitive to small discriminations in distance. And indeed the map furnishes the reader with the preferred post office of a goodly proportion of the area's farmers. Aylmer, containing post office no. 2, drew custom from farmers immediately surrounding it as well as eastwards along the major country road leading to it, thus elongating the eastern portion of its "market" area. Luton, with post office no. 25, displays also an elongated service area, skewed heavily toward the east and south of it. Curiously, Luton could not attract the loyalty of Mr. James McCausland whose farm abuts the hamlet on its northern margin (he preferred to trek to Aylmer three miles away!), whereas Mr. A. Miller, a mile down the road closer to Aylmer, must have encountered McCausland often when heading in the opposite direction to pick up his mail in Luton.

Beyond the question of mail, the pattern of schools suggests a fairly orderly dispersion of school houses to cater to a dispersed child population; likewise with the symbols that denote mills (★), blacksmiths' shops (crossed circles), and cheese factories (open circles). More irregular, however, is the distribution of country churches. No less than three churches were strung out at mile intervals along the road east of Aylmer. These may seem minor patterns, but they add up to the means by which communities defined their membership, trade centers carved out territories that stimulated growth or not, and the general scale and pace of life in an orderly Canadian rural district such as this developed.

CHAPTER NINE

Two Examples of Thematic Maps: Civil War and Fire Insurance Maps

Robert Karrow and Ronald E. Grim

*M*ost of the maps that we have been considering until now have been general geographical maps of one kind or another, more concerned to show the general appearance of the land than to concentrate on any particular spatial phenomenon. During the nineteenth century, though, there was a huge increase in the production of what are called *thematic* maps, which deal with some particular theme. Such subjects might be demographic patterns, the spread of railroads, the location of military posts, the nature of buildings in a city, and so forth.

It would obviously not be possible to examine the whole range of maps of this kind. Moreover, some of them have good guides that mention both their historical potential and their location: books like Peter J. Guthorn's *United States Coastal Charts 1783–1861* (Exton, Pa.: 1984) or Andrew Modelski's *Railroad Maps of North America: The First Hundred Years* (Washington, D.C.: 1984).

However, there are two categories of thematic maps that offer dramatic possibilities to the historian, and yet have not been widely used: these are the military maps of the Civil War, and fire insurance maps. The military maps of the Civil War cover an area of the country that had been relatively neglected in large-scale maps. The Wheeler Survey (Chapter 10) was active in the southwest; much of the northeastern and middle states, as well as the midwest, had been mapped by commercial firms in landownership maps and atlases (Chapter 8). But the southeast had not been included in these cartographic ventures, and some of the most detailed nineteenth-century maps of that area were drawn during operations of the Civil War. At the time of Reconstruction, the federal government launched a project to collect and print all the pertinent records of the "War of the Rebellion," as it was called, and maps were included among these records.

The *Official Records* of the Union and Confederate Armies filled 128 volumes and were supplemented by 178 atlas plates. These were originally issued in thirty-seven parts, but are usually found bound into one, two, or three volumes with the title *Atlas to Accompany the*

Official Records of the Union and Confederate Armies. Most of the atlas plates are subdivided into smaller maps, so that there are fully 775 separate maps in the *Official Atlas*. They are of two types: battle maps and military campaign maps. A third kind of map, ground plans of military forts and installations, was considered to be of primarily architectural interest and so has not been included here.

The battle maps tend to be at a large scale, sometimes as large as one inch to one thousand feet, and to show the location and movements of individual fighting units. Campaign maps tend to be at a smaller scale, from about one inch to one mile, to one inch to sixteen miles, and so function as general topographic maps of the country. Four-fifths of the maps come from the Union side, and all were drawn under widely differing circumstances, sometimes relying on careful compass traverses or plane-table surveys, but sometimes having been drawn hastily during a cavalry reconnaissance. In each case, the map reproduced in the *Official Atlas* bears a note indicating the source or sources; a number of them, for instance, are by the great Confederate topographer Jedediah Hotchkiss.

The arrangement of maps in the *Official Atlas* roughly parallels the arrangement of the *Official Records* themselves, and is essentially chronological. The maps are basically redrawings of maps that accompanied the original reports, but there are many discrepancies; maps were fitted in because they were the right size to complete a page makeup, or they turned up too late in the editing process to be incorporated into their correct sequence. In any case, there is no geographical sense to the arrangement and one can find maps of Kentucky, Louisiana, and Virginia on the same page, so that the *Graphic Index* cited in the bibliography to this chapter is an indispensable finding aid.

Most of the maps use multiple colors—black, brown, blue, green, and red all being employed—which results in a very attractive appearance. Because of its military importance, relief tends to be carefully drawn, usually with hachures but sometimes, on the largest scale maps, using contours (Plate 9.4). Wooded areas are often shown with a green tint. Because of the great variety of scales, there are considerable differences in the amount of detail recorded on the maps. All show roads, railways, rivers, and settlements, and most show relief, forested areas, and swamps. Most larger-scale maps show street patterns in cities and towns, individual buildings in rural areas, and also bridges, fords, churches, and schools. Many maps (almost one half) show the names of landowners or residents, making the *Official Atlas* a valuable supplement to the maps and atlases described in Chapter 8. The larger-scale campaign maps often identify specific industrial sites such as distilleries, limekilns, paper mills, woolen mills, icehouses, quarries, mines, ironworks, and so forth. Very large-scale maps may show fences or cultivated fields. Plans of cities and towns show and often name streets and public buildings, and many show individual houses.

In this latter respect the information contained on the *Official Atlas* maps sometimes supplements the material found in our second category of thematic map, the fire insurance map. These large-scale urban maps were originally created as a reference tool for fire insurance underwriters, and gained wide acceptance after the middle of the nineteenth century. By the beginning of the twentieth century, when the Sanborn Map Company in the United States and the Charles E. Goad Company in Canada had become the dominant producers, their coverage had been extended to most communities of any size in the two countries. The Sanborn Map Company's production reached its peak in the early 1930s, and during its almost one hundred years of activity the Company covered more than 13,000 U.S. towns.

Typical fire insurance maps are hand-colored lithographs, normally printed on sheets measuring twenty-one by twenty-five inches, with a scale of one inch to fifty feet. Coverage for small communities was issued on several loose sheets, while that for the largest cities ran to several bound volumes. Because the surveys were so detailed, and the production runs small, the maps were quite expensive. They were not sold to individual customers, but rather issued on a subscription basis, with the Sanborn Map Company recalling outdated volumes when new editions were prepared, or replacing corrected pages, or applying paste-on revisions as construction changes were noted.

These maps show a wealth of information about the physical characteristics of individual buildings and fire-fighting apparatus, using color coding, symbols, and abbreviations. The basic plan shows the street pattern and street widths, as well as the location and shape of individual buildings. Color coding is used to identify a building's construction material (pink—brick, yellow—frame, blue—stone, green—iron, and brown—adobe), while a variety of symbols are used for roof composition. Abbreviations record the use of buildings ("D" for dwelling, "D.G." for dry goods, "F" for flat, "F.B." for female boarding, and so forth), while the names of major commercial and industrial establishments are indicated on the map, as well as house numbers, heights of buildings, and number of stories.

Although numerous firms were involved in the production of fire insurance maps, their content and format were quite similar, no doubt because of the early adoption of standard map symbols under the dominance of the Sanborn Map Company. The Sanborn system as it eventually developed is explained in the *Surveyors' Manual for the Exclusive Use and Guidance of Employees of the Sanborn Map Company* (New York: 1950; reissued in 1911, 1923, 1925, and 1936). There is an explanation of the standard map legend and a building-by-building interpretation of several sample blocks in the last edition of the *Catalog of Insurance Maps* (New York: 1950). Although map legends appear on each set of maps, this publication is a useful resource for interpreting the content of Sanborn maps.

Fire insurance maps appeal to a wide range of researchers. Chadwyck-Healey, Inc., in its promotional literature for selling microfilm copies of the Library of Congress holdings, enumerates a long list of potential users: genealogists, local historians, urban historians, geographers, demographers, economists, and so forth. It would be futile to attempt to list all the potential users, but the value of this source is becoming increasingly apparent both to map custodians and to historical researchers, as our section on "work using fire insurance maps" will show. In summary, the depiction of individual buildings and the identification of their use on these maps make them a priceless resource for reconstructing urban landscapes during the late nineteenth and early twentieth centuries.

SOURCES FOR CIVIL WAR MAPS AND FIRE INSURANCE MAPS

AVAILABILITY OF THE OFFICIAL ATLAS

Most large and medium-sized public libraries will own either the original issue of the *Official Atlas* or a reprint edition; they will be cataloged under "U.S. WAR DEPT." The number for the atlas volumes in the Superintendent of Documents classification is W45.7, and they may also be found as part of the Congressional Serial Set, having been issued as House Miscellaneous Document 291 in the 52nd Congress, 1st Session.

The first forty plates were reprinted by a commercial publisher, Atlas Publishing Co. of New York, in 1892 as *Atlas of the War of the Rebellion*. In 1958 Thomas Yoseloff published a full reprint with an introduction by Henry Steele Commager under the title *The Official Atlas of the Civil War*, while the most recent reprint (slightly reduced in size) is by the Fairfax Press (New York: 1983) under the title *The Official Military Atlas of the Civil War*.

STUDIES USING FIRE INSURANCE MAPS

Applebaum, William. "A Technique for Constructing a Population and Urban Land Use Map." *Economic Geography* XXVIII (1952), pp. 240–43. Explains the use of fire insurance maps in reconstructing past patterns of population, urban morphology, and land use.

Aspinall, P. J. "The Use of Nineteenth-Century Fire Insurance Plans for the Urban Historian." *The Local Historian* XI (180), pp. 342–49. Interesting discussion of the use of these plans in the British context.

Bloomfield, G. T. "Canadian Fire Insurance Plans and Industrial Archeology." *IA, The Journal of the Society for Industrial Archeology* VII (1982), pp. 67–80. This article opens up a theme and some research possibilities that have barely been touched in the United States.

Bowden, Martyn J. "Downtown through Time: Delimitation, Expansion and Internal Growth." *Economic Geography* XLVII (1971), pp. 121–35. An example of the use of fire insurance maps in the analysis of urban change through time.

Conzen, Kathleen Neils. "Mapping Manuscript Census Data for Nineteenth-Century Cities." *Historical Geography Newsletter* IV (1974), pp. 1–7. Shows how to use city directories and fire insurance maps to map manuscript census data.

Gibson, Lay James. "Tucson's Evolving Commercial Base, 1883–1914: A Map Analysis." *Historical Geography Newsletter* V (1975), pp. 10–17. Another example of the use of fire insurance maps to examine urban change.

Lamb, Robert B. "The Sanborn Map, a Tool for the Geographer." *California Geographer* II (1961), pp. 19–22. A wide-ranging discussion of the uses of fire insurance maps.

Merrimack Valley Textile Museum. *A Checklist of Prints, Drawings and Paintings in the Merrimack Valley Textile Museum.* North Andover, Mass.: 1973. Under "Mills and factories" are lists of fire insurance maps for the New England states and New York State.

Ross, Stanley H. "The Central Business District of Mexico City as indicated on the Sanborn Maps of 1906." *Professional Geographer* XXIII (1971), pp. 31–39. An illustration of the use of these maps in Latin America.

Sauder, Robert A. "The Use of Sanborn Maps in Reconstructing 'Geographies of the Past,' Boston's Waterfront from 1867 to 1972." *Journal of Geography* LXXIX (1980), 204–213.

Wright, Helena. "Insurance Mapping and Industrial Archeology." *IA, The Journal of the Society for Industrial Archeology* IX (1983), pp. 1–18. An excellent analysis of the nature of these maps, of their availability, and of their research potential.

Wrigley, Robert L. "The Sanborn Map as a Source of Land Use Information for City Planners." *Land Economics* XXV (1949), pp. 216–19. One of the earliest articles to draw attention to the uses of these maps.

AIDS IN USING THE OFFICIAL ATLAS

Aimone, Alan C. *The Official Records of the American Civil War: A Researcher's Guide.* 2nd ed. West Point, N.Y.: 1977. This is a good, brief introduction to the entire series of *Official Records.* Aimone points out some pitfalls for the user resulting from the nature of the compilation and editing.

Hotchkiss, Jedediah. *Make Me a Map of the Valley: The Civil War Journal of Stonewall Jackson's topographer.* Ed. Archie P. McDonald. Dallas: 1973. An account of how maps were actually drawn at this period.

Irvine, Dallas. *Military Operations of the Civil War: A Guide-Index to the Official Records of the Union and Confederate Armies, 1861–1865.* 5 vols. Washington, D.C.: 1861–1865. A detailed account of the records that attempts to supplement their inadequate index.

LeGear, Clara E. *The Hotchkiss Map Collection: A List of Manuscript Maps, Many of the Civil War Period, Prepared by Major Jed. Hotchkiss.* Falls Church, Va.: 1977. A reprint of the edition first published by the Library of Congress in 1951.

O'Reilly, Noel S.; David C. Bosse; and Robert W. Karrow, Jr., eds. *Civil War Maps: A Graphic Index to the Official Atlas.* Chicago: 1987. An atlas, arranged alphabetically by state, to show where the maps are to be found in the *Official Atlas.*

SOURCES FOR FIRE INSURANCE MAPS

By far the largest list is contained in *Fire Insurance Maps in the Library of Congress: Plans of North Ameri-*

can *Cities and Towns Produced by the Sanborn Map Company,* prepared by the Geography and Map Division at the Library of Congress and published in Washington, D.C., in 1981. These maps have been microfilmed by Chadwyck-Healey, Inc., of 1021 Prince Street, Alexandria, Virginia 22314, from whom details of availability and price may be requested. Some other listings of this material include:

Curtis, Peter H., et al. *Fire Insurance Maps of Iowa Cities and Towns: A List of Holdings.* Iowa City: 1983.

Hoehn, R. Philip, et al. *Union List of Sanborn Fire Insurance Maps Held by Institutions in the United States and Canada.* 2 vols. Santa Cruz, 1976–1977.

Karrow, Robert W., Jr., ed. *Checklist of Printed Maps of the Middle West to 1900.* 14 vols. in 12. Boston: 1981–1983. Much information concerning Sanborn maps in the twelve midwestern states covered by this cartobibliography.

Rees, Gary, and Hoeber, Mary. *A Catalogue of Sanborn Atlases at California State University, Northridge.* Santa Cruz: 1973.

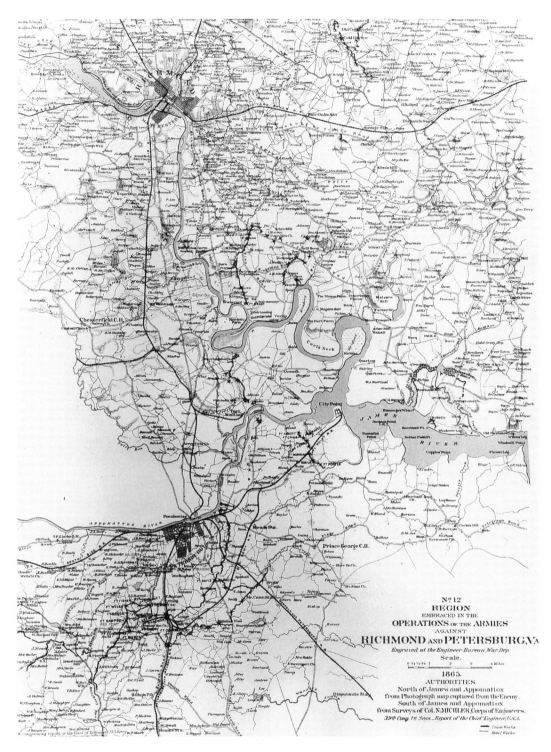

Plate 9.1. "Region embraced in the operations of the armies against Richmond and Petersburg," from the *Official Atlas* of the Civil War.

Plate 9.1a. Modern map to show area of Plate 9.1.

A Detailed View of the Virginia Countryside, c. 1865

This plate in the *Official Atlas* of the Civil War is called "Region embraced in the operations of the armies against Richmond and Petersburg [Virginia]" and the information comes from various sources. North of the James and Appomattox Rivers it is taken from a captured Confederate "photograph map," and south of them it comes from a survey made by the (Union) Corps of Engineers in 1865. In its general appearance and level of detail this is typical of the medium-scale campaign maps in the *Atlas*.

Purely military information is limited to the rings of fortifications around Richmond and Petersburg, the military railroad from City Point to Petersburg, and various other scattered fortifications. Three kinds of road are shown: turnpikes by double lines, country roads by single lines, and poor roads or trails by dotted lines. The profusion of personal names attached to tiny black squares indicating dwelling houses is reminiscent of the eastern style of land ownership maps and atlases. The names could be a primary source for genealogical research, and the number and location of houses and barns could support a study of population density or of agricultural production.

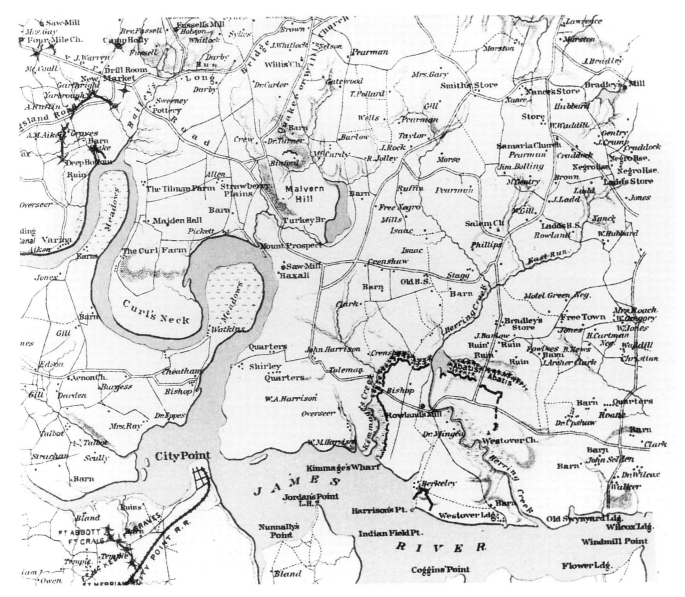

Plate 9.2. Detail from
Plate 9.1.

A Closer Detail of the James River Valley

This enlargement of the same map focuses on an important stretch of the James River Valley. This section of the river, about thirty-five miles upstream from the site of the Jamestown colony, and thirty-five miles below the fall line at Richmond, marked the effective head of navigation. City Point, at the southwest corner of the map, served as the "deep water" port for Petersburg, supplies making the rest of the journey into town by rail. City Point became Grant's headquarters and supply depot during his sieges of Petersburg and Richmond between June of 1864 and April of 1865.

Seven landings and wharves are shown on the river below City Point, three of them, on the north side of the river, serving two of the most important Virginia plantations. Westover was founded in 1691 by William Byrd I (1652–1704), and became the seat of a famous Virginia family still active in politics; Westover Landing and Church mark it on this map. To the west of Westover is Berkeley, home of the Harrison family that sired two presidents. It was founded by Benjamin Harrison III (1673–1710), and beside Berkeley itself and Harrison's Point, we can see the residences of John, W. A., and W. M. Harrison.

Within the ten-mile square shown by this enlarged detail (perhaps 10 percent of it water), we can count almost 100 names of residents, three stores, two grist mills, and a sawmill, a pottery, two blacksmith shops (abbreviated "BS"), two churches, a number of isolated barns (tobacco warehouses or curing sheds?), and four "ruins," presumably casualties of war. Slavery is prominently in evidence, for an overseer's house is marked on the Harrison estate, and in three locations are rows of buildings marked "quarters." The residences of six blacks, all presumably free, are shown individually. One is labeled "Free Negro," three "Negro Hse.," and two others have the abbreviation "Neg." following the personal name. In addition, there is a "Free Town," apparently a small settlement of free blacks, just a little north of Westover.

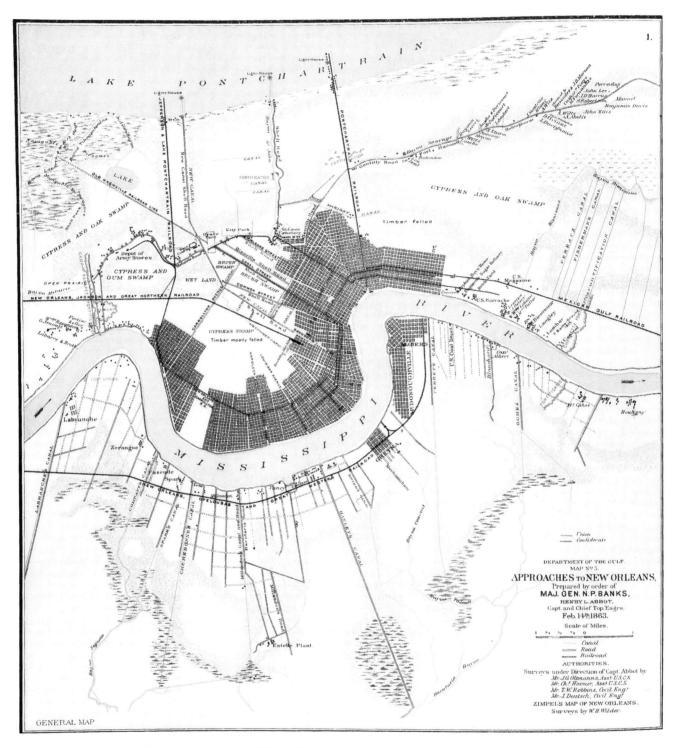

Plate 9.3. "Approaches to New Orleans," from the *Official Atlas*.

New Orleans and Its Surroundings in 1863

This map is based on Charles Zimpel's great plan of the city, published in 1834 and measuring roughly five feet square. Needless to say, the Zimpel plan is very rare except in the larger libraries of Louisiana. There are some military features on this map, especially the Confederate lines of fortification. But it contains as well a wealth of information about the city and its surroundings.

The most obvious change from the plan of 1720 (Plate 3.6) lies in the great extension of the built-up area, following the rather clumsy street lines at right angles to the river. Railroad lines now serve the city on both the north and the south banks of the river, and there is a road below the levee. The outline of the rural estates follows the long-lot pattern shown on Plate 4.6 and on Color Plates 1 and 2, and there are many names of owners. On some of the plantations the outline of the buildings may be seen; at "Labranche," for instance, on the extreme left of our plate, we can identify the distinctive slave quarters.

There are now many canals, and two railroads sticking out rather oddly into Lake Pontchartrain. Close examination will reveal various industrial establishments such as sugar refineries and brickyards, though the swamp and bush still hem the city in closely. Perhaps the area in the "U" of the city marked "Cypress swamp, timber mostly felled" indicates the shape of things to come, for after a period of stagnation after the war, New Orleans would show growth extensive enough to encompass this whole area and beyond.

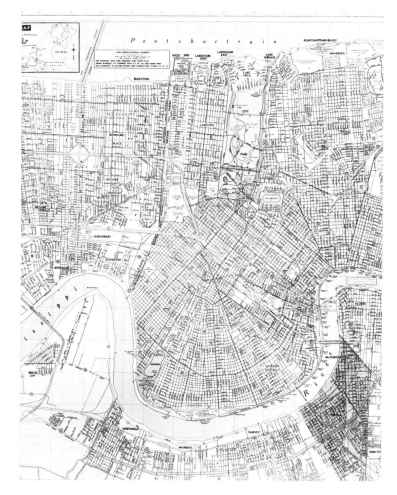

Plate 9.3a. Detail showing the city from *Metropolitan New Orleans Tour Guide Map*, published by the Gulf Oil Company (1973 edition). Reproduced courtesy of Chevron Corporation.

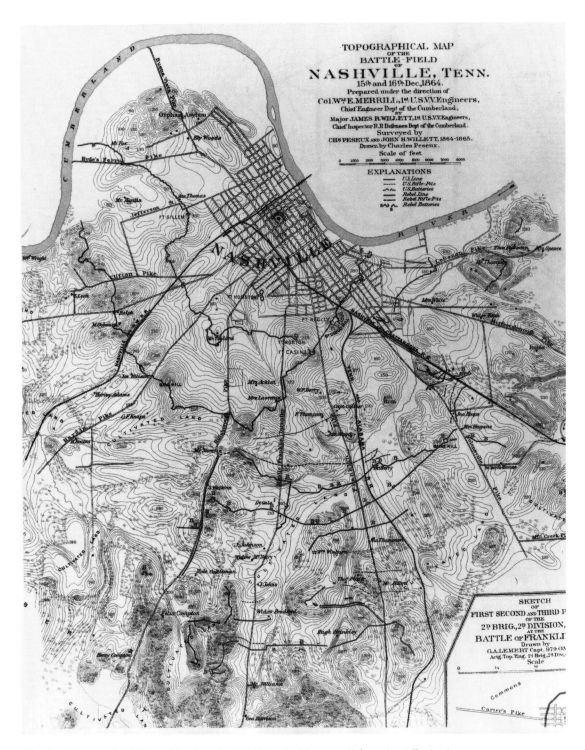

Plate 9.4. Topographical Map of the Battlefield of Nashville, Tennessee," from the *Official Atlas*.

Plate 9.4a. Detail showing Nashville from *Nashville and Vicinity,* published by the Exxon Oil Company (1974 edition). Map courtesy of General Drafting Company, Inc.

Nashville and Its Surroundings in 1864

Unlike the map in Plate 9.3, this map is an example of work prepared entirely by the military engineers. There are many military features, in particular the Union lines around the south of the city, and the Confederate posts and batteries beyond them. But there are also numerous elements allowing us to reconstruct Nashville's civilian appearance.

The map is closely netted with contour lines, giving us an impression of this broken country much more accurate than had been possible with hachures or shading like the work on Plates 1.11 and 7.2. The roads have been inserted with care, and so have the railroad lines; many names of landowners, and some of institutions, also appear. There is also some indication of "cultivated land"; all in all, this map would be very profitable to compare with a USGS topographical map of the same area.

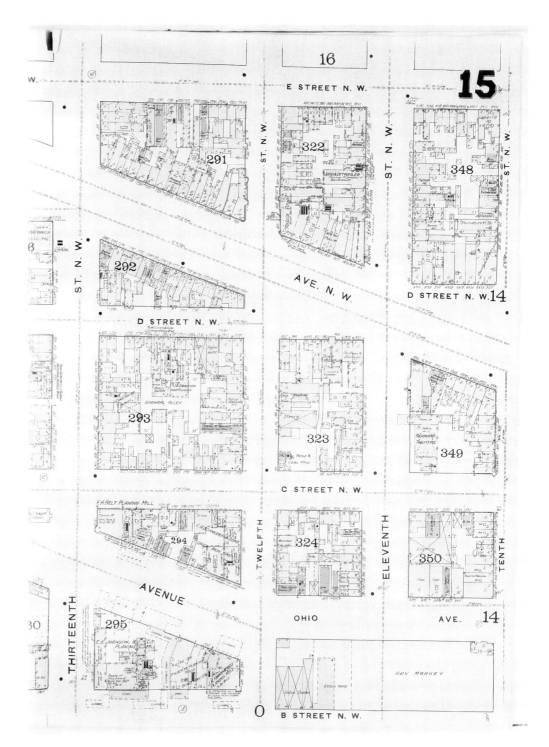

Plate 9.5. Detail from *Insurance Maps of Washington,* by the Sanborn Map Publishing Company (New York, 1888).

Washington's Federal Triangle Area in 1888

One of the primary research possibilities of fire insurance maps is to be able to study change through time in one area of a town. The example used in the next three plates is an area in downtown Washington, D.C., bounded on the north by E Street, on the south by Constitution Avenue, and on the east and west by 10th and 13th Streets, N.W. This is part of the area known today as the Federal Triangle, primarily containing federal government office buildings. However, it has not always served that function, and by looking at plans from three time periods, it is possible to follow the dramatic changes that have occurred in this area.

In 1888 this part of town was characterized by a mixture of commercial and light industrial buildings. Some of the more prominent structures included Kernan's Theater (block 349), the Evening Star Building (a daily newspaper, block 322), and Hay Market. There were also numerous small stores (indicated by an "S"), most of which were two or three stories high and constructed of brick. Other services and industries that can be identified include hotels, boardinghouses, printing shops, furniture warehouses, auction and storage facilities, carpenters' shops, planing mills, machine shops, foundries, livery and food stores, and wood and coal yards. One can almost picture the activity around the planing mills in block 295, where the store of lumber on the adjoining streets is indicated.

The map contains much technical information of primary interest to fire fighters, but also includes house numbers (a three- or four-digit number on the street side of each building), the number of stories (the number in the street-front corner of a building), and the building's height. Of course it also sets out the original layout of the streets, which would soon be heavily modified.

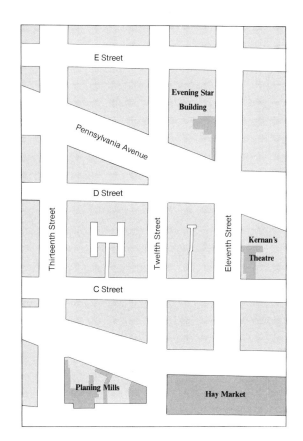

Plate 9.5a. Diagram of Plate 9.5.

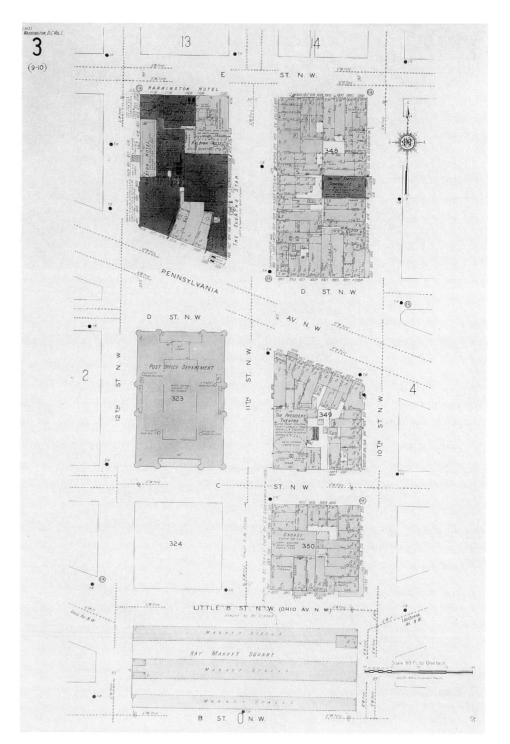

Plate 9.6. Detail from
Insurance Maps of Washington,
by the Sanborn Map
Publishing Company (New
York, 1928).

Changing Land Use in the Federal Triangle Area, 1928

This 1928 map documents the area immediately prior to the major construction programs conducted by the federal government during the 1930s. The most dramatic changes are occurring in blocks 322, 323, and 324. Block 322 experienced considerable consolidation of lots, as three major buildings replaced the small two- and three-story brick stores. The three buildings were constructed in stages reflecting the gradual acquisition of adjoining lots, the Raleigh Hotel from 1897 to 1912, the Harrington Hotel in 1916 and 1926, and the Evening Star Building in 1900 and 1921.

Even more dramatic is the complete removal of all previous buildings in blocks 323 and 234; 324 remains vacant, while on 323 is the building occupied by the Post Office Department. It had eight stories, and a 150-foot clock tower, making it a major landmark on Pennsylvania Avenue. The adjoining three blocks (348, 349, and 350) experienced much less change. Many of the three-story brick stores remain, though several new buildings have appeared. The livery in block 350 has been replaced by a garage with a fifty-car capacity, and the feed stores have gone, but nearby is an automobile repair shop; signs of the times. Hay Market, with its market stalls oriented in an east-west direction, remains a prominent feature.

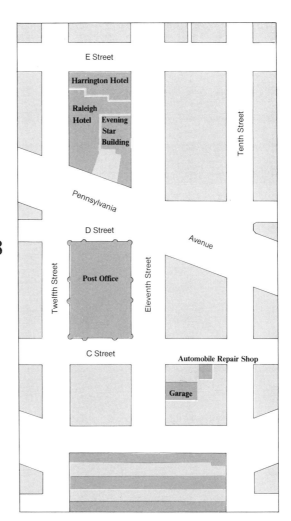

Plate 9.6a. Diagram of Plate 9.6.

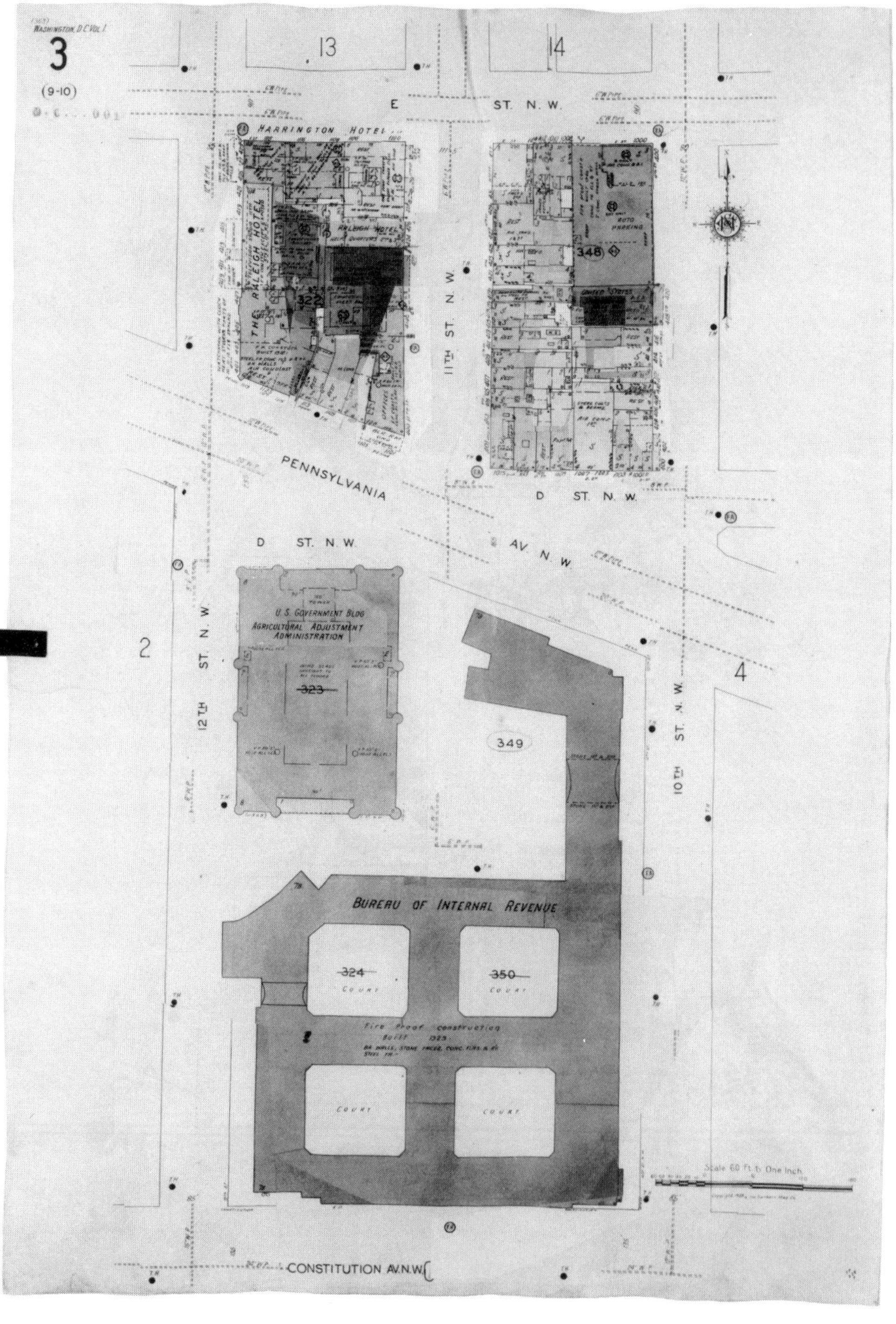

Plate 9.7. Detail from *Insurance Maps of Washington,* by the Sanborn Map Publishing Company (New York, up to 1959).

The Emergence of the Federal Triangle, c. 1960

Although it may not be immediately apparent from our plate, this map of the same area provides an example of the paste-on corrections that the Sanborn Company employed to reduce production costs. Using a 1928 base map, thirty-one editions of paste-on corrections were issued by the company up to 1959. Although the atlas from which this plate was taken records changes made up to 1959, there is no key to identifying the most recent changes on this particular plate.

The most significant changes recorded here are in the area south of Pennsylvania Avenue. Blocks 324, 349, and 350, as well as Hay Market, have been replaced by one of the Federal Triangle government office buildings, the Internal Revenue Service. The original design for the Federal Triangle called for the removal of the Post Office Building in block 323, which at the time of the last correction on this map was occupied by the Agricultural Adjustment Administration rather than by the post office. Having survived demolition, though, the "Old Post Office," as it is now known, was renovated in 1983. At present, the ground floors house restaurants and boutiques, while the upper floors provide office space for several government agencies, including the National Endowment for the Humanities, partial sponsor of this manual.

The two blocks (322 and 348) north of Pennsylvania Avenue have changed the least. The Raleigh and Harrington Hotels and the Evening Star Building still dominate block 322, while block 348 is still composed of small brick stores. This was in 1959; today all the buildings in this block, except for the facades on the block's northwestern corner and the U.S. Storage Company, have been leveled, making way for a new commercial office complex. By 1988, a hundred years after the first example used here, almost all the buildings shown on the 1888 map in this six-block area will have been replaced.

Source: Frederick Gutheim and Wilcomb E. Washburn, The Federal City: Plans and Realities *(Washington, D.C.: 1976).*

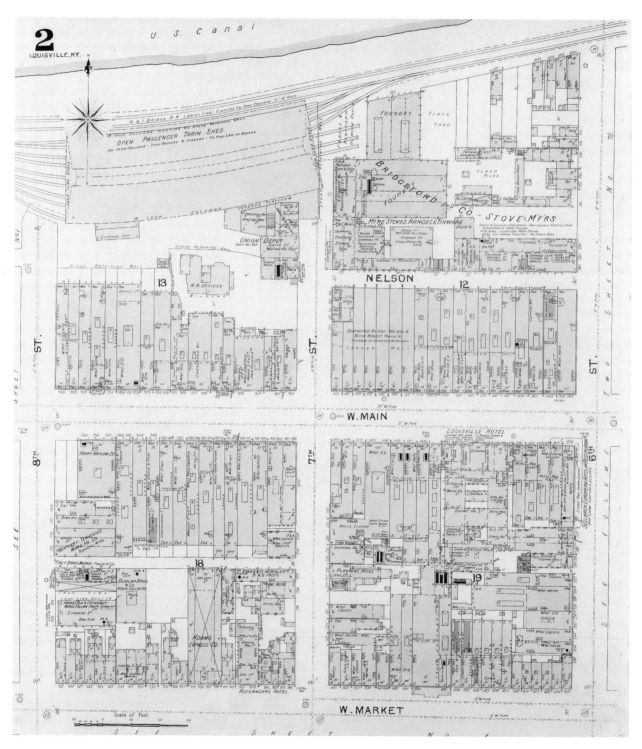

Plate 9.8. Detail from *Insurance Maps of Louisville, Kentucky*, by the Sanborn Map Publishing Company (New York, 1892).

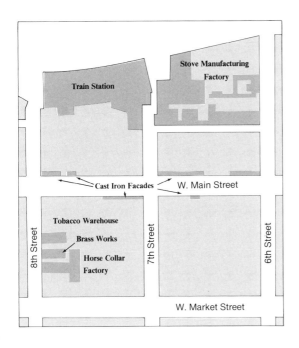

Plate 9.8a. Diagram
of Plate 9.8.

Wholesaling in Downtown Louisville, 1892

One way of using fire insurance atlases is to take slices of the same area of a city at different periods; another is to take slices of different parts of the same city at the same period. Here we shall analyze three different parts of Louisville in 1892. According to the 1890 census, Louisville at that time had a population a little over 160,000, making it the twentieth largest city in the United States. It had been founded in 1778 as a fort at the falls of the Ohio River, and had grown as a center of communications, whether by canoe, barge, or railroad, eventually developing as well such industries as tobacco processing, distilling, meat packing, shipbuilding, textile manufacture, flour milling and food processing.

Our plate shows a four-block area of downtown Louisville, mainly a wholesaling district. A great variety of goods are for sale, and the train station in the northern half of block 13 provides evidence of the city's importance as a transportation center. The same area has six hotels, and several industries such as a stove manufacturing factory, a horse collar factory, a brass works, and a tobacco warehouse.

Almost all the buildings are of brick with four or five stories, though a number of the wholesaling establishments on Main Street (601–611, 623–643, and 729–745 on the north side, and 640–642 and 700–726 on the south side) also have cast-iron facades. Many of these buildings are still standing today; and this collection of cast-iron facades, which is listed on the National Register of Historic Places, is reputed to be the best collection of such facades outside of New York City.

Source: Ronald M. Greenberg and Sarah A. Marusin, eds.,
The National Register of Historic Places, 1976
(Washington, D.C.: 1976).

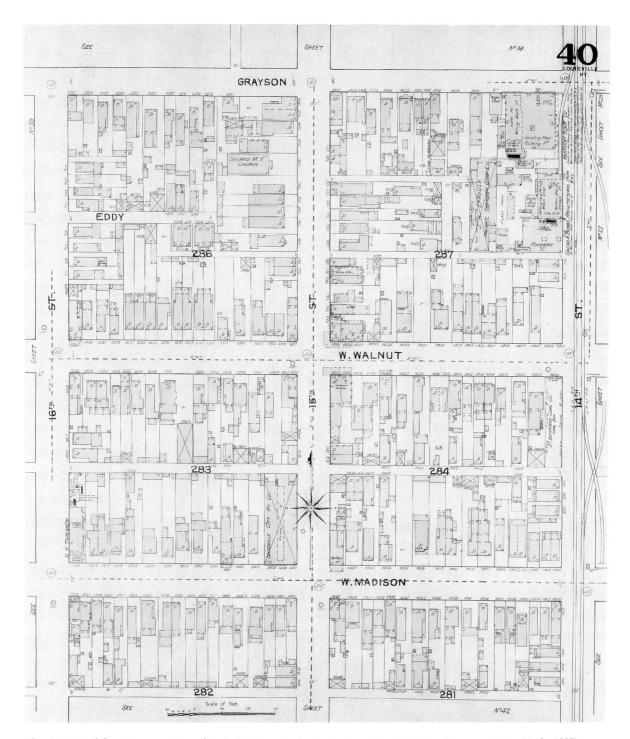

Plate 9.9. Detail from *Insurance Maps of Louisville, Kentucky*, by the Sanborn Map Publishing Company (New York, 1892).

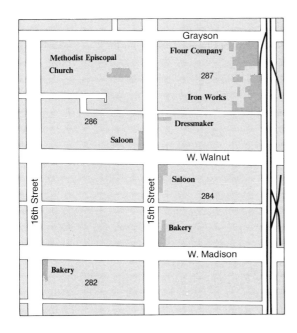

Plate 9.9a. Diagram of Plate 9.9.

Residential Patterns in Louisville, 1892

This plate shows a residential area that is about three quarters of a mile southwest of the previous downtown example. The majority of these structures are dwellings (indicated by "D") and most of them one- or two-story frame buildings. However, about half of the houses fronting on Walnut Street as well as the four stores ("S") and the two saloons ("sal.") at the intersection of 15th and Walnut Streets are constructed of brick. There are several in-house businesses including a dressmaker in block 287 and bakeries in blocks 282 and 284. Larger industrial activities include a flour company and ironworks in block 287. The railroad along 14th Street also reminds us of Louisville's importance as a center of communications.

Of particular interest is the colored Methodist Episcopal Church in block 286, which suggests that the ethnic identification of religious and educational facilities provides a key to locating specific ethnic populations. For instance, other map sheets covering neighborhood blocks show the West End Colored Public School on Chestnut Street between 15th and 16th Streets, while on 12th Street there is another colored Methodist Episcopal Church. A correlation of the house numbers recorded on the map with those listed in the 1892 City Directory indicates that the residents of all the houses on Eddy Street (or alley) are listed as colored, while the residents of all the houses in block 286 fronting on Walnut, Grayson, 15th, and 16th Streets are not. However, a few residents in the 1400 blocks of Grayson and Walnut are also listed as colored. This apparently dispersed pattern conforms to the general notion that there were no large concentrations of blacks in Louisville in the late nineteenth century; rather, their residences were intermixed with those of the whites.

Source: John L. Anderson, "Changing Patterns of Louisville's Black Residential Areas," in An Introduction to the Louisville Region: Selected Essays, *ed. Don E. Bierman (Louisville: 1980).*

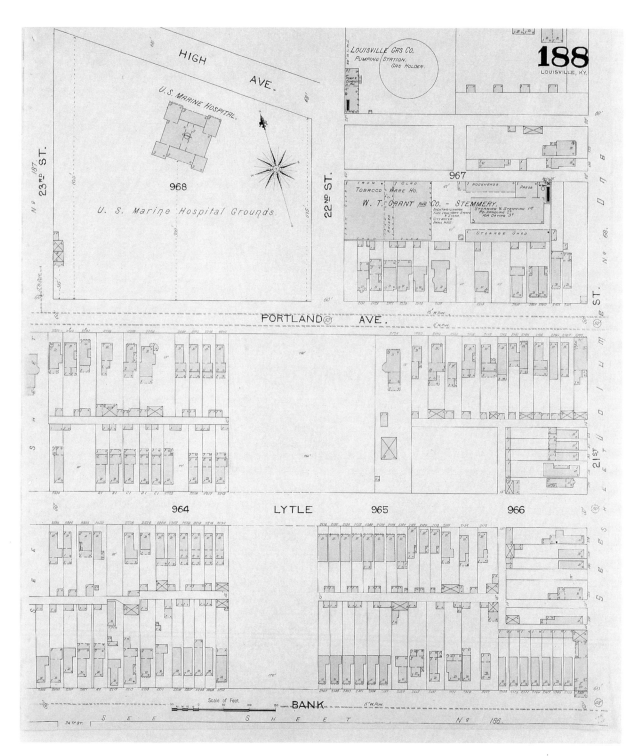

Plate 9.10. Detail from *Insurance Maps of Louisville, Kentucky,* by the Sanborn Map Publishing Company (New York, 1892).

More Residential Patterns in Louisville, 1892

The third example depicts another residential area, about a mile and a half northwest of the first downtown example. These several blocks are part of the Portland community, founded in 1814 as a separate town at the lower end of the falls of the Ohio River. Although Portland was incorporated into Louisville in 1852, its original town plan is still evident. Portland's streets are laid out in a grid pattern that parallels the southeast to northwest orientation of the Ohio River, so that it does not coincide with the Louisville grid, aligned in a north-south, east-west direction.

Again, residential dwellings are the predominant feature on the map, but these houses are a special type: they are the one-room–wide cottages known as *shotguns*. Many of the shotguns have only one story, although there is a variety called the *camelback* that has two stories in the rear. Built in the 1850s, many of the remaining shotguns are today being restored. This plate also reminds us of Louisville's industrial activity, in the shape of the W. T. Grant and Company's tobacco warehouse and stemmery, and of her vocation as a communications center, for the U.S. Marine Hospital was built by the federal government in 1847 for the care of sick and injured rivermen.

Source: A. William Dakan et al., The Dainty Guide: A Louisville Handbook *(Louisville: 1980)*.

CHAPTER TEN

Topographic Surveys of the United States

Robert Karrow

As a general rule, the usefulness of a map for historical purposes varies directly with the scale, maps at the largest available scales (those showing a smaller area in more detail) generally being more useful. That is not only because they are more detailed, but also because they are likely to be more accurate, having been based on actual measurements made on the ground.

The ideal situation would be to have complete coverage of a country at a large scale, with systematic revision, so that changes over time could easily be compared. Such comparisons are possible for much of western Europe since the early nineteenth century. In England, for instance, the successive versions of the Ordnance Survey maps (see Chapter 1) cover the country with a series of quadrangles, the sheet lines of which are regular divisions of latitude and longitude. By 1873 the whole country had been mapped at a scale of an inch to a mile, in 110 sheets, and subsequent revisions of these maps offer remarkable sources for the study of modern British history.

In the United States, the lack of an early, decisive entry by the federal government into large-scale topographic mapping meant that such activities were fragmented and unco-ordinated, a situation that prevailed for the first hundred years of the republic. Furthermore, the enormous size of the country dictated a smaller scale map in the early stages, for if it had been mapped at one inch to a mile, using sheets the same size as those of the British Ordnance Survey, some 13,400 sheets would have been required. At the scale that eventually emerged as the standard topographical scale of the United States Geological Survey (USGS), roughly two and a half inches to the mile coverage of the coterminous United States takes some 53,600 sheets. It is only the advent of aerial photography and automated mapping techniques that has made it possible to almost complete the coverage at this scale today.

Little wonder then, given the magnitude of the task, that the systematic topographic mapping of this country got off to a slow start, or that the scales chosen were often much smaller than an inch to a mile. The first proposal for a systematic map of the United States at a uniform scale, with one inch equaling ten miles, was made by the ingenious engineer and inventor Christopher Colles in 1794. His *Geographical Ledger* proposed a plan for a "united collection of topographical maps," each sheet covering an area 2 degrees of latitude by 4 degrees of longitude.

Colles developed this system to cover the entire globe, but only five sheets were ever produced, showing most of New England and New York and parts of Pennsylvania and New Jersey. The proposal was visionary and its fulfillment beyond the means of the chronically poor Colles. Such a project needed government support, but Colles's scheme for a universal topographic map was universally ignored. Another world map at the scale of one inch to ten miles was proposed by Colonel Henry James of the British War Office in 1859, but his results were not much more substantial than Colles's; only twenty-four published sheets are known, eighteen covering eastern North America.

The next essay at a systematic map of the United States met with more success. In 1869, George M. Wheeler, a young lieutenant in the U.S. Army, proposed a uniform topographic map of the area west of the 100th meridian of longitude. His system, as originally proposed, called for ninety-five sheets, each covering 1 degree 40 minutes of latitude and 2 degrees 45 minutes of longitude, at the scale of one inch to eight miles. Wheeler, with the aid and support of General A. A. Humphrey, Chief of the Corps of Engineers, managed to organize, launch, and direct a survey that operated for ten field seasons.

In 1879 the U.S. Geological Survey was formed, and funding for the Wheeler survey was cut off. But despite an unsympathetic Congress and the active hostility of two other governmental surveys in the West (neither of which was concerned with systematic mapping) the Wheeler survey managed to publish fifteen sheets at the scale of one inch to eight miles and another thirty-four "quarter sheets" at the scale of one inch to four miles. These sheets, taken together, cover a good part of the

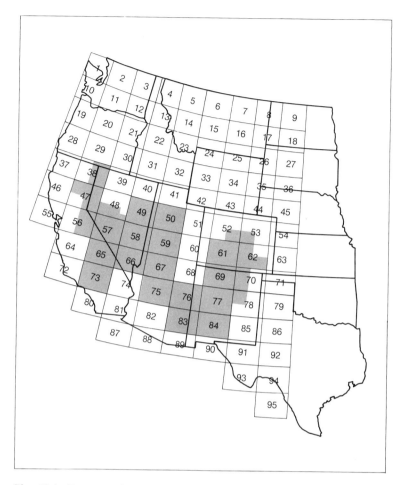

Plate 10.1. Coverage of the topographic maps published by the "U.S. Geographical Surveys West of the One-hundredth Meridian," Lt. George M. Wheeler in charge.

area originally intended for mapping (see Plate 10.1), and they can be used as a topographical and historical benchmark for the decade of the 1870s (Plates 10.3 and 10.4).

The maps were intended primarily to serve military purposes such as transportation, planning, and logistics, but they were also used as base maps on which to plot geological and land use data. They show a grid of latitude and longitude at 30-degree intervals, relief by hachures with occasional spot elevations, water bodies and rivers, springs, marshes, sand, and alkaline flats. Cultural information includes cities, towns, and isolated settlements, Army forts and camps, railroads, wagon roads, trails over grades impractical for wagons, telegraph lines, and state boundaries. Place names are a class of cultural feature that may be taken for granted on maps, but should not be; it is remarkable how often maps of the same area differ in their toponymy. Maps of the Wheeler survey are likely to be of special value in place-name studies, since they are based on field work and so reflect actual local usage.

The cultural features enumerated above are those shown with special symbols on the legend sheet accompanying the *Topographical Atlas* and may be assumed to be consistently mapped. Many other cultural features indicated by name on individual sheets may not be identified consistently throughout the *Atlas*. These include ranches, abandoned forts, toll houses, sawmills, brickyards, post offices, Indian settlements, and so forth. Some of the published maps of the survey were also overprinted to show geological formations or information about land use. These maps contain the same base information as the topographical series, but may be rendered more difficult to read by the overprinting.

With the establishment of the USGS in 1879, systematic mapping by the Wheeler survey ceased, and the Department of the Interior became the sole agency concerned with the survey of U.S. lands. One of its arms, the General Land Office, was responsible for producing the cadastral map of federal lands, as we have discussed in Chapter 4. The other agency, the Geological Survey, had as its primary goal the publication of a uniform topographical map of the country. Although these USGS maps are not yet complete at every scale for the whole country, they do constitute the basic map source for most historical studies; hence their frequent use in other chapters of this manual (Plates 7.8, 8.4a, 12.7a, 12.8a, 12.9a, 12.10a, and 12.11a).

The first topographic sheets of the USGS appeared in the early 1880s, but progress was slow. Much of the survey's resources went into geological investigations and reports and special geological mapping, and the size of the country conspired against rapid completion of the maps. Furthermore, the topographical mapping of the USGS has usually been done on a cooperative basis, with the individual states providing part of the funds, and some states have encouraged mapping more than others.

Sheet lines are based on latitude and longitude. At various times, there have been quadrangles *(quads)* of 1 degree, 30 minutes, 15 minutes, and 7.5 minutes on a side, with scales ranging from 1:250,000 (1 inch equal to about 4 miles) for the 1 degree quad to 1:24,000 (2.5 inches to the mile) for the 7.5-minute quad. In the early years of the survey, the smaller scales were used more often. From about 1910 to 1950 the 15-minute quad (1:63,360 or 1 inch to the mile) dominated production and seemed likely to be the standard map of the survey. Since 1960, however, there has been a pronounced shift to the 7.5-minute quad at 1:24,000 or, more recently, a 1:25,000 metric scale. The production of USGS quads is summarized in Plate 10.2.

There are, then, four different scales for USGS maps, but complete coverage of the United States is currently available only at the smallest of these scales, 1:250,000. At the time of writing there is a commitment to complete the coverage at 1:100,000 for use as base maps for the 1990 census, but in order to meet this deadline, many sheets are being produced in planimetric editions only, without contour lines. Although a greater number of points and greater detail can naturally be shown on the larger-scale maps, the whole range of USGS maps shows much the same type of data. Relief is their hallmark, using contour lines that follow lines of equal elevation with intervals of from five to twenty feet. Spot heights and the elevation of benchmarks are frequently shown,

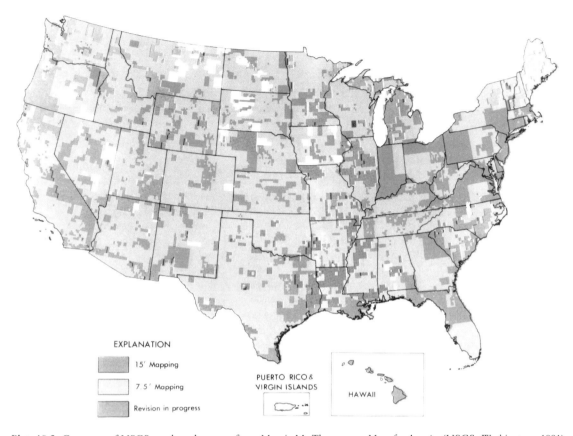

EXPLANATION

15' Mapping

7.5' Mapping

Revision in progress

PUERTO RICO &
VIRGIN ISLANDS

HAWAII

Plate 10.2. Coverage of USGS quadrangle maps, from Morris M. Thompson, *Maps for America* (USGS, Washington, 1981).

and there is a highly accurate grid of latitude and longitude that makes it possible to refer precisely to any point on the map, a feature used by archeologists for recording the locations of sites.

The full range of civil boundaries is shown, as well as the lines of the township and range survey where this was used. Earlier sheets were printed in three colors: brown for contour lines, black for cultural features, and blue for hydrography. A very wide range of cultural features is shown, and in the late 1940s green overprinting began to be used to show forested areas. In short, these maps offer a detailed analysis of any area for which they exist, and when a sequence of them exists in time, they offer a

chronology of the region that it is hard to establish in any other way.

The topographical map can help bring vividly to mind the physical contexts in which North American history has been played out. Michillimackinac and the route of the Oregon Trail can be studied on the topographic map, as can the bloody hills south of Gettysburg and the mill sites of Lowell. Much of the topography on a modern map can be taken to represent the ground surface at an earlier data, though the user must be aware of possible changes. Bunker Hill is no longer evident on the ground, and Mount Juliet, a prominent hill shown on many eighteenth-century maps of Illinois, had been mined away for building materials by the time of the Civil War. Hydrography is similarly mutable: lakes are drained, reservoirs formed, swamps are drained, rivers change course or dry up.

Still, much of our "historical topography" has been but little altered, though it may take some concentration to see beneath the mass of cultural information on a topographic map to the land surface below. Because of the increasing density of cultural features and the changes they may impose on landforms, it is indeed often the earliest USGS map in a series that is the most suggestive. Topography, cultural features, toponymy, and change over time can all be studied in these marvelous maps, as the examples will show.

SOURCES FOR THE TOPOGRAPHIC SURVEY MAPS

SOURCES OF WHEELER SURVEY MAPS

The published maps and reports of the Wheeler Survey were widely distributed, at least in the early years of the survey, and most college libraries were sent sheets as they were published. When cataloged in North American libraries, survey materials will be entered under the official name of the survey as "U.S. Geographical Surveys West of the 100th Meridian." If they have been retained as part of a U.S. Government Documents collection, the maps and reports may bear the Superintendent of Documents classification number "W8." The final reports and maps of the Wheeler Survey were not issued as part of the Congressional Survey set, although the survey's annual reports are included in the report of the Chief of Engineers.

The single most complete collection of the published maps of the survey is probably that in the Geography and Map Division of the Library of Congress. The National Archives holds, in addition to a collection of published sheets, all of the manuscript maps and field books from which the printed maps were produced. There are also a number of incomplete manuscript drawings for sheets that were never published. The bibliography is not substantial:

Bartlett, Richard A. *Great Surveys of the American West*. Norman: 1962. This reference contains two chapters on the Wheeler Survey.
Goetzmann, William H. *Exploration and Empire: The Explorer and the Scientist in the Winning of the American West*. New York: 1966. This book contains a chapter on the Wheeler Survey.
————. "The Wheeler Surveys and the Decline of Army Explorations in the West." In *The American West: An Appraisal,* ed. Robert G. Ferris. Santa Fe: 1963.

Thompson, Gilbert. "Bibliographical List of Maps and Atlas Sheets Published by the Office of Explorations West of the 100th Meridian." In *A List of Geographical Atlases in the Library of Congress,* vol. I, ed. P. Lee Phillips. 1909, pp. 705–13. This is a complete listing of the sheets, including many variant printings, in the Library of Congress collection. This list refers to sheets by number; sheet numbers are shown on the index map (Plate 10.1).

SOURCES OF USGS MAPS

Thanks to a depository program of long standing, many libraries in the United States have complete, or virtually complete, sets of USGS quadrangle maps. Older and larger institutions can generally be expected to have a full set of quads for the whole country from the earliest to the most current, while smaller libraries may have only current sheets, or may be only partial depositories, acquiring quads of only one state or region.

Current USGS quads can be purchased in person or through the mail. For maps of areas east of the Mississippi, including Minnesota, the address is:

Branch of Distribution
U.S. Geological Survey
1200 South Eads Street
Arlington VA 22202

For maps of areas west of the Mississippi, including Louisiana, it is:

Branch of Distribution
U.S. Geological Survey
Box 25286, Federal Center
Denver CO 80225

Maps of local areas are also available over the counter from nine USGS Public Inquiries Offices, located in Anchorage, Los Angeles, San Francisco, Menlo Park (Calif.), Denver, Washington, Dallas, Salt Lake City, and Reston (Va.). In addition, maps of local areas can be purchased (with a retail markup) from some 1,650 commercial dealers; the authorized dealers for a state are listed in the index to maps of that state.

Out-of-print quadrangles can be obtained in photographic reproduction from the National Cartographic Information Center (NCIC), a branch of the Geological Survey. The address is:

NCIC—U.S. Geological Survey
507 National Center
Reston VA 22092

The NCIC can also supply out-of-print and current maps on 35 mm microfilm. Sheets published between 1884 and 1972 constitute the "Historical File" or "A series," and film is available by states, on reels containing about 500 quads each.

Also available from the NCIC, free of charge, is the appropriate state "Index to Topographic Maps." This is the basic tool for identifying quadrangle maps, and consists of a state map overprinted with a grid to show which quads are available, and their names. Quads are specified by giving the state name, series, and quad name, for example, KANSAS, 7½ MINUTE, LAKIN SW QUAD.

As sheets are replaced by revised editions or go out of print, the older editions no longer appear on the indexes. This can create a problem for the historian attempting to locate the earliest (or some intermediate) USGS map of a given area. The historical map file on microfilm, mentioned above, has all the quads arranged in alphabetical order, but since no old index maps were reproduced as part of the file, a researcher could have great difficulty in discovering that a given area of interest was first mapped in, say, a thirty-minute quad of 1908.

Such problems are neatly solved by the retrospective graphic index to USGS quads, 1882–1940, compiled by Moffatt (see his bibliography listing). Moffatt's compilation effectively replaces similar kinds of graphic indexes prepared for Wisconsin and the North Central States (see the bibliography listings under "Galneder" and "Karrow").

BIBLIOGRAPHY FOR USGS MAPS

Carrington, David K., and Richard W. Stephenson. *Map Collections in the United States and Canada: A Directory*. 4th ed. New York: 1985.

Cobb, David A. *Guide to U.S. Map Resources*. Chicago: 1986.

Galneder, Mary, et al. *A Union List of Topographic Maps of Wisconsin (1:24,000 to 1:1,000,000)*. Madison: 1975.

Karrow, Robert W., Jr., ed. *Checklist of Printed Maps of the Middle West to 1900*. 14 vols. in 12. Boston and Chicago: 1981–1983. Indexes pre-1901 USGS quads of twelve midwestern states.

Makower, Joel, and Laura Bergheim. *The Map Catalog; Every Kind of Map and Chart on Earth and Even Some Above It*. New York: 1986. An excellent compendium of topographical and other maps, with information about their availability.

Moffat, Riley M. *Map Index to Topographic Quadrangles of the United States, 1882–1940*. Santa Cruz: 1986.

Thompson, Morris M. *Maps for America: The Cartographic Products of the U.S. Geological Survey and others*. 2nd ed. Reston, Va.: 1981. Excellent survey of the range of USGS production, illustrated by hundreds of color reproductions of details from quads.

Plate 10.3. Detail from sheet 49 of the Wheeler Survey.

Plate 10.3a. Modern map to show area of Plate 10.3.

The Eureka Mining District at the Height of Its Prosperity

This detail comes from sheet forty-nine of Wheeler's survey, covering east central Nevada. It was one of the earliest sheets published, and was based on field work done in 1869 and 1872. It is substantially accurate, and shows the Eureka and White Pine Mining Districts at the height of their booms.

The Eureka Mining District was organized in 1864 after the discovery of rich deposits of lead and silver ores. The town of Eureka was established in 1869, and grew to a peak of nine thousand inhabitants in 1878. Its sixteen smelters earned it the nickname "the Pittsburgh of the West," and one of the largest, the Richmond Consolidated Smelter, is indicated south of the town. Although Eureka survives today as a county seat with a population of five hundred, the neighbouring towns of Pinto and Vanderbilt have vanished; indeed, Vanderbilt township was abolished as early as 1876.

In the White Pine Mining District to the east, the size of lettering symbolizes the importance of Hamilton, which was platted in 1868 and by the next year was home to ten thousand people. Another ten thousand were scattered in other towns and mining camps around Hamilton; the area had become the scene of the biggest mining rush since California in 1849. Much of Hamilton burned in 1873, while nearby Treasure City had already lost most of its six thousand people by 1870; it burned in 1874. Wheeler's map thus captures the district at the peak of its brief day in the sun.

The map shows an abundance of mines and mills, as well as a variety of transportation routes. It also brings out the general north-south pattern of relief, which made communications difficult. In fact, the whole complex was dependent on its communications with the outside world, for all food and other supplies had to be freighted in, and the venture was kept afloat by the shipments of ore and bullion, which in the summer reached one hundred tons daily.

Sources: W. Turrentine Jackson, Treasure Hill: Portrait of a Silver Mining Camp (Tucson: 1963); Nell Murbarger, Ghosts of the Glory Trail: Intimate Glimpses into the Past and Present of 275 Western Ghosttowns (Las Vegas: 1983); and Stanley Paher, Nevada Ghost Towns and Mining Camps (Berkeley: 1970).

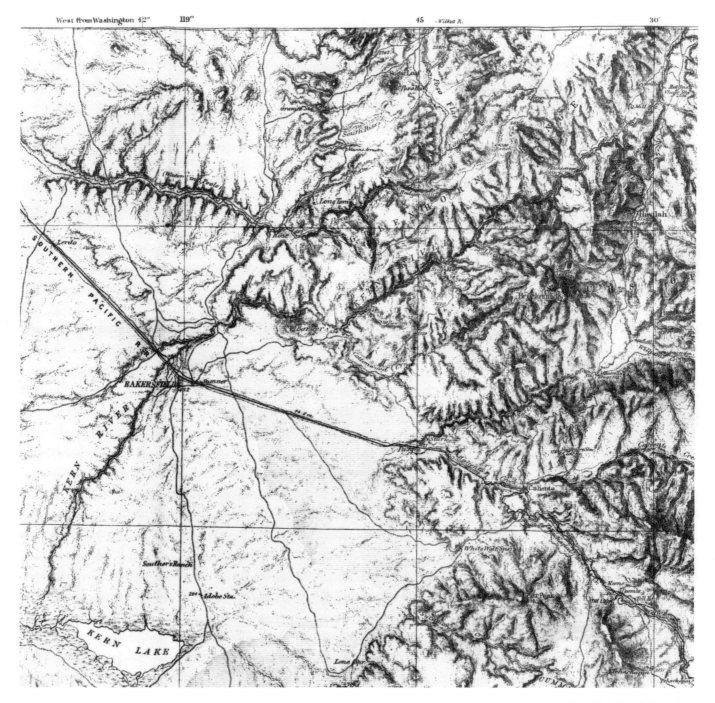

Plate 10.4. Detail from sheet 37 of the Wheeler Survey.

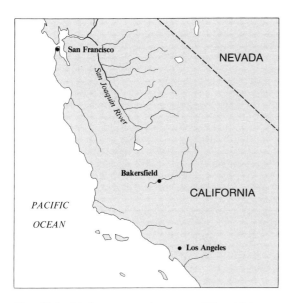

Plate 10.4a. Modern map to show area of Plate 10.4.

Site of the First Gold Rush in Southern California

This detail is from one of the last sheets of the Wheeler survey to appear (sheet 37). It is based on surveys made in 1875 and 1878 and was published in June 1879. Comparison with Plate 10.3 demonstrates the one advantage that a topographic series will always have over separately conceived and published maps, however detailed: their relative uniformity over space and time. Not only do the sheets of a topographic series employ the same scales and symbols and have the same overall look, but we are usually justified in assuming that they show the same density and types of information.

The Southern Pacific Railroad had been completed to Caliente, near the bottom of the detail, in April of 1875. The very steep section beyond Caliente to the southeast, through Tehachapi Pass, had been completed by the time Wheeler's party returned in 1878. The route is the one recommended by Lts. Williamson and Parke in their report on the Pacific Railroad surveys of 1853–1854. Bakersfield was platted in 1866 and three years later had a population of 600. The 1880 census recorded 1,000, so Wheeler's map presumably shows a town of about 750 people. The railroad was originally supposed to run through the town, but citizens balked at the Southern Pacific's demand to be granted a right-of-way two blocks wide, so the route was shifted a mile northward and a new town, Sumner (now East Bakersfield) emerged.

To the northeast of Bakersfield, in the southern spurs of the Sierra Nevada, gold was discovered near Keysville in 1854 and soon led to the first gold rush in Southern California. Los Angeles merchants were delighted at the prospect of a boom at their end of the state, but although only 110 air miles separated Los Angeles from the diggings, formidable mountains and the Mojave Desert intervened. The decade of the 1860s saw many roads built through the rugged terrain south of the Kern River, roads that contemporaries described as the worst they had ever seen. It is hard to pick out these roads in the mountainous terrain on our photograph, but they are there, and are even distinguished between wagon roads, trails, and trails "impracticable for wagons."

Many of the Wheeler maps were available in an edition overprinted in colors to show land use, one of the earliest appearances of this technique in the United States (Color Plate 5). On this sheet the area south of Bakersfield is colored to represent "Agricultural [lands] (with irrigation)," and in fact numerous canals and ditches built around the turn of the century have made it prime farming land. Areas of tillable land also appear in mountain valleys and basins, while chapparal and timber are indicated higher up the slopes. The distribution of wooded areas on this sheet is strikingly similar to the modern forest cover shown on the USGS 1:250,000 map (Bakersfield quad) covering the same area.

Source: Eugene Burmeister, The Golden Empire: Kern County California *(Beverly Hills: 1977).*

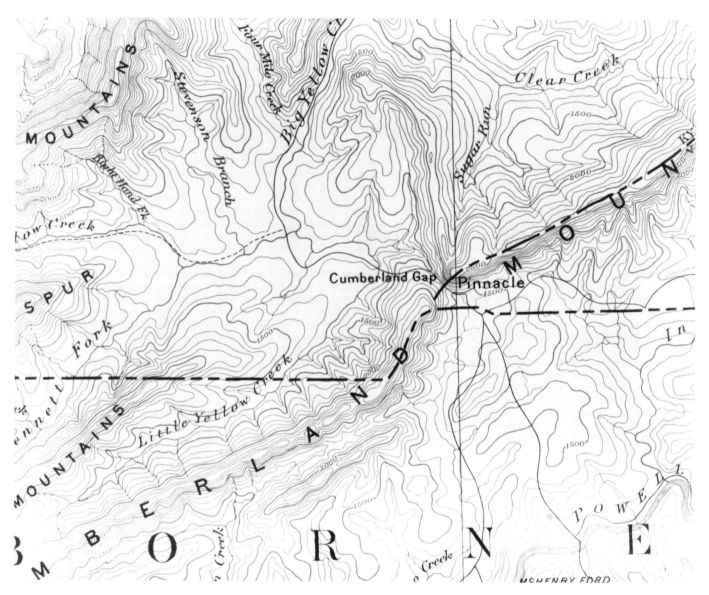

Plate 10.5. Detail from the USGS 30-minute Cumberland Gap, Virginia quad (1882 edition).

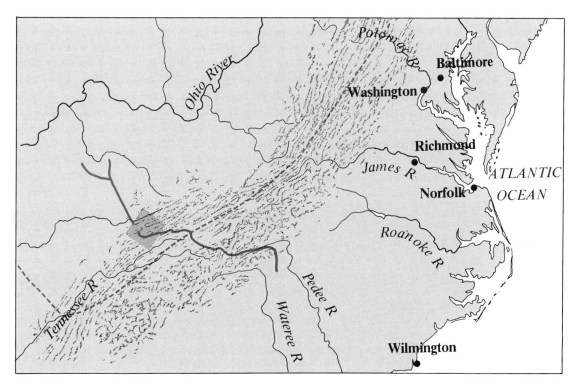

Plate 10.5a. Modern map to show location of the Cumberland Gap.

A Nineteenth-Century View of a Strategic Gap

Cumberland Gap is a natural break in the Cumberland Mountains at the western tip of Virginia, and had been traversed by Indians for thousands of years before Dr. Thomas Walker entered the Kentucky country through it in 1750. Almost twenty years later John Finley and Daniel Boone passed through, and Boone returned to settle in Kentucky in 1775. He was followed by waves of settlers in the first great overland migration in U.S. history, with as many as twenty thousand people going through the Gap in one year.

The relatively small-scale 30-minute Cumberland Gap quad reproduced here is quite extraordinary in presenting the area virtually as it would have appeared in 1750 or, for that matter, in 1550. There are no railroads, no settle-ments, almost no cultural features at all except the wagon road through the Gap. Surveyed in 1882, this was one of the first USGS quads, and the Geological Survey's reports of iron and coal deposits encouraged an infusion of settlers. The town of Middlesboro was platted west of the Gap, at the junctions of two trails and two rivers, in 1889, and a railroad was constructed through the Gap in a mile-long tunnel in 1891; these features will appear on Plate 10–6.

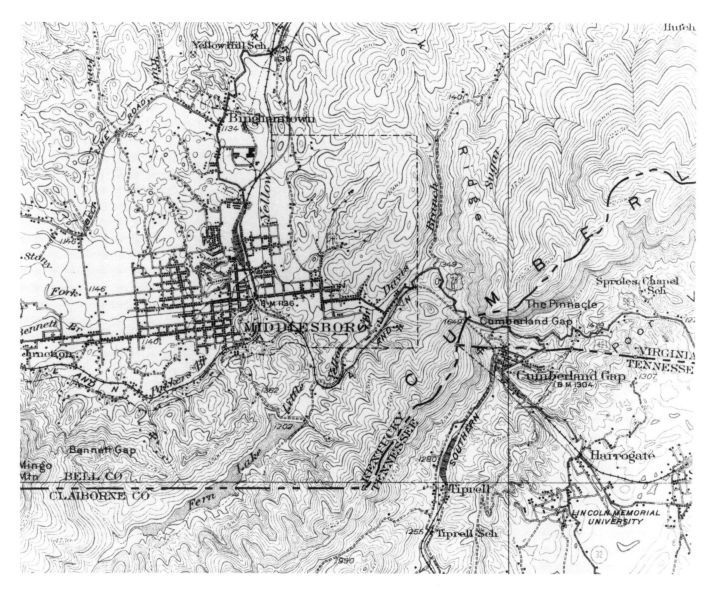

Plate 10.6. Detail from the USGS 15-minute Middlesboro quad, Virginia (1926–27 edition).

Survivals and Developments around the Gap

This is an extract from the 15-minute Middlesboro quad surveyed in 1926–1927. The land on both sides of the Gap has now been settled, but the modern transportation routes tend to follow the original tracks. Thus the U.S. Highway follows the course of the "Old Wilderness Road," and the Indian trail (the so-called Warrior's Path) that preceded it.

The Gap itself remains instantly identifiable, with its walls rising from 800 to 1,200 feet above the valley floor at the east end, and the narrowest point on the floor of the Gap being only about 100 feet wide. Of course, the Gap played its part not only in the westward expansion, but also in the Civil War. At the top of "The Pinnacle," on the north end of the Gap, is the site of Fort Lyons, a point so important tactically that it changed hands four times during that war.

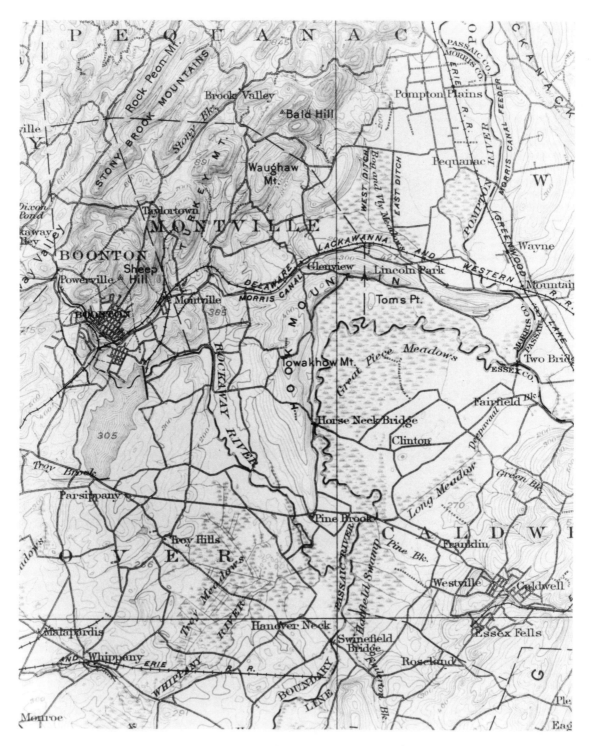

Plate 10.7. Detail from the USGS 15-minute Passaic quad, New Jersey (1887–1903 edition).

On the Track of Rochambeau's Eighteenth-Century Army

This quad shows part of north central New Jersey, and brings out the way in which old routes may often be found to underly modern roads and even interstate highways. In 1783 the army of Rochambeau marched from Pompton to Whippany, one leg of the road from Newport to Yorktown. The French engineers left detailed itineraries and maps of their routes, and these allow us to establish them with considerable certainty.

The contemporary French map shown here (Plate 10.7a) has been marked with letters, beginning at "A" and ending at "G," with corresponding letters on the Passaic quad. Rochambeau's engineers described the country in the text, and on their map the coincidence with the modern roads is clear, even though the French map takes the form of a strip rather than an areal sketch. In the next example of a USGS map (Plate 10–8), we shall look more closely at one section of this itinerary.

Source: Howard C. Rice, Jr., and Anne S. K. Brown, The American Campaigns of Rochambeau's Army; 1780, 1781, 1782, 1783, *2 vols. (Princeton/Providence: 1972).*

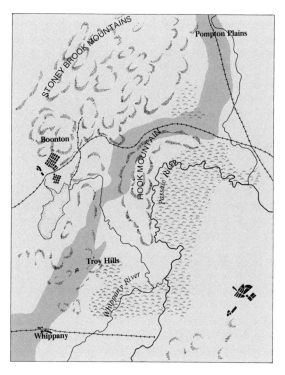

Plate 10.7b. Modern map to show area of Plate 10.7.

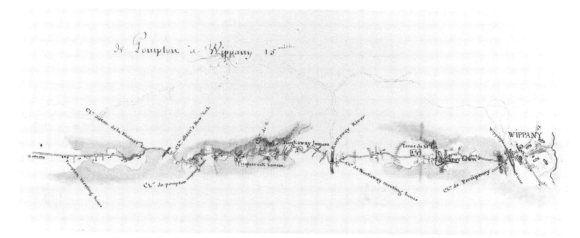

Plate 10.7a. Detail from *The American Campaigns of Rochambeau's Army,* ed. Howard Rice and Anne Brown (Princeton, 1972). Papers of Louis Alexandre Berthier. Princeton University Library.

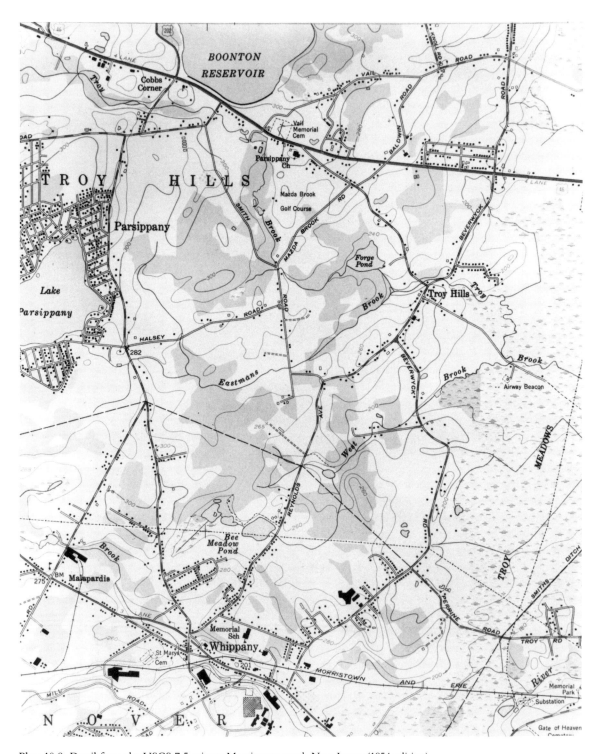

Plate 10.8. Detail from the USGS 7.5-minute Morristown quad, New Jersey (1954 edition).

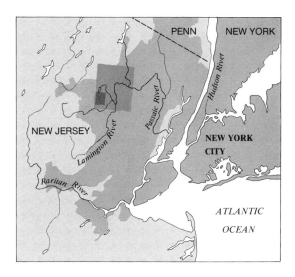

Plate 10.8a. Modern map to show location of Plate 10.8.

Rochambeau's Route in the 1950s

This map shows the last few stages on Rochambeau's journey from Pompton to Whippany, and is marked as on the French map shown as Plate 10.7a. There are, of course, many modern intrusions such as golf courses and airline beacons. Still, taking a text like the one that Rochambeau's engineers noted down at "E," we read that "the country is open on the left and there are woods quite close on the right; you come to a crossroads where you intersect at right angles a highway from Morristown to New York."

At this intersection there is now a shopping center, but the line of the old roads, and the lay of the land, are perfectly recognizeable. It would be interesting to check the date of some of the buildings along Reynolds Avenue and Beverwyck Road, since some may well date from the time of the Revolution. The place-names on a map constitute a special class of cultural feature that is among the most enduring, though subject to corruption in transmission from one language to another, or in copying from one map to another. Here, for instance, Rochambeau's "Troy Town" survives as Troy Hills, "Percipenny" as Parsippany, and "Piquareck" as Pequanec.

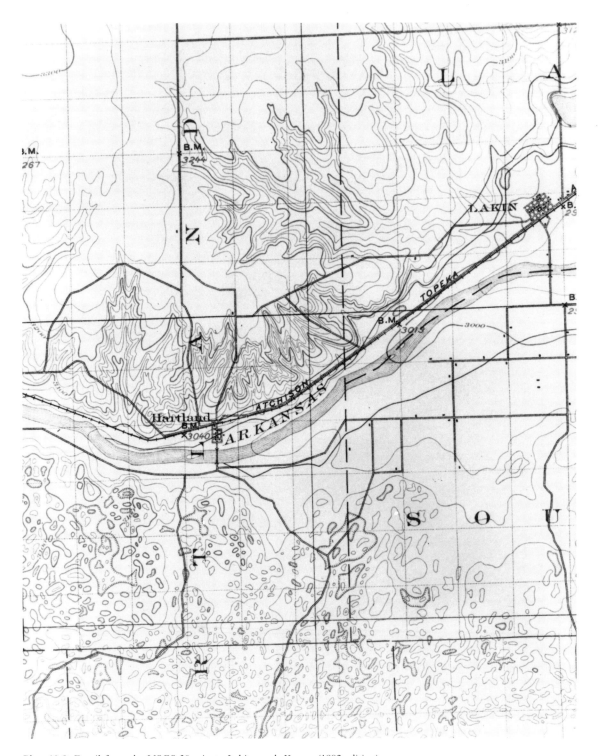

Plate 10.9. Detail from the USGS 30-minute Lakin quad, Kansas (1892 edition).

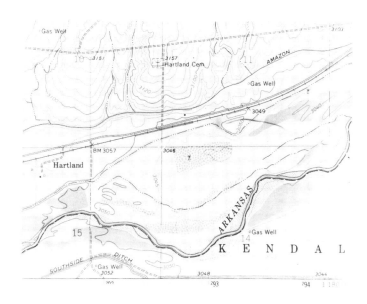

Plate 10.10. Detail from the USGS 7.5-minute Lakin quad, Kansas (1966 edition).

The Decline and Fall of the Community of Hartland, Kansas

In 1887 the two-year-old town of Hartland, in Kearney County, Kansas, was a thriving community with two newspapers, four hotels, brass band, three restaurants, six general stores, and an ice cream parlor. The population exceeded 1,000 and it was considered the leading commercial center of southwestern Kansas. It lost a county seat battle to nearby Lakin in 1896 and its decline thereafter was rapid. On the first USGS map to cover the area, the 30-minute Lakin quad surveyed in 1892 and shown here in detail, Hartland is already only a shadow of its former glory, with nine streets, about a dozen buildings, a railroad station, and a quarter-mile-long bridge across the Arkansas River (the bridge washed out in 1901).

Hartland continued to decline. The last business, a store, burned in 1927, and the post office was closed in 1933. One will now look in vain for it on most modern maps of Kansas, even in the *Rand McNally Commercial Atlas and Marketing Guide*, usually the last bastion of neglected flag stops and abandoned railroad sidings. Still, it survives as a name on the 1966 7.5-minute quad, where it now refers only to a house and two outbuildings at the end of a dirt road on the north bank of the Arkansas River. If these buildings one day disappear, and "Hartland" is no longer used even to designate an intersection, the name will probably be perpetuated by the cemetery northeast of the old townsite.

Good as these maps are for toponymic studies, the student does not necessarily have to go at once to them. The Geological Survey has been working for some years on a series of state gazetteers that will index all the names, including such microtoponyms as the names of schools, churches, and cemeteries, appearing on all quads for that state. Lists for certain states are now available on microfiche and as computer printouts, and formal published versions for some states have started to appear.

Source: History of Kearney Co., Kansas, *vol. 1 (Lakin, Kan.: Kearney County Historical Society, 1964).*

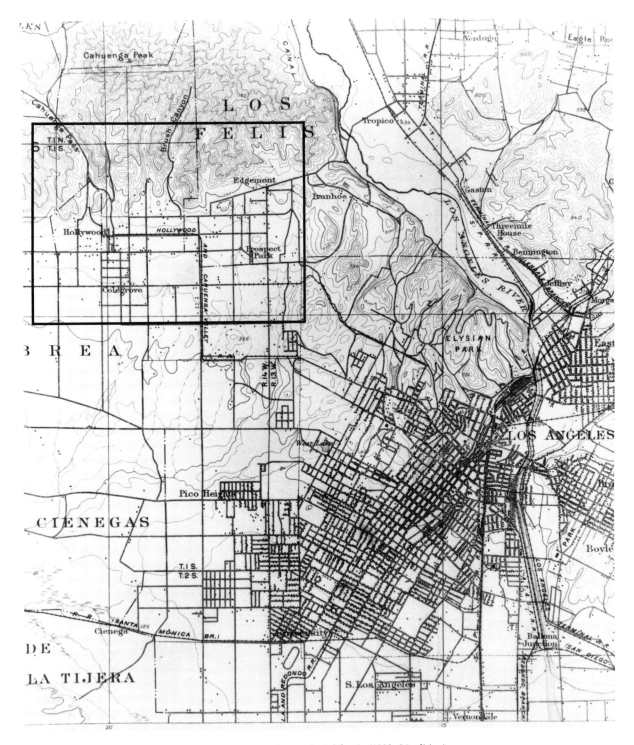

Plate 10.11. Detail from the USGS 15-minute Los Angeles quad, California (1892–94 edition).

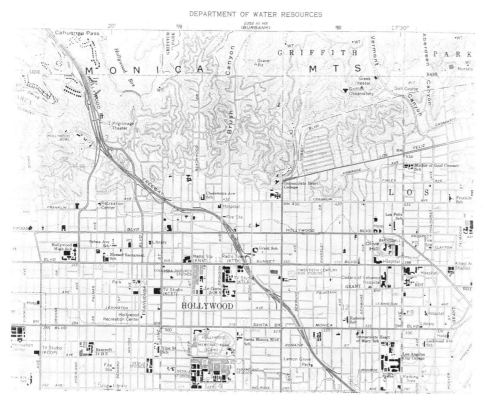

Plate 10.12. Detail of the USGS 7.5-minute Hollywood quad, California (1966 edition).

From Spanish Ranch to Tinseltown

As with fire insurance maps, so with USGS quads; it is sequences in time that are particularly revealing for city development. Here we see a detail from the 1893–1894 survey of the Los Angeles area. The city, with a population of about 50,000, has expanded from its historic core (the solidly lined blocks) only as far southwest as "University," the site of the infant University of Southern California. Beyond that is a region of, presumably, truck farms and ranches spreading west to Santa Monica.

The old Spanish grants of "Las Cienegas," "La Brea," and so forth are shown prominently. The village of Hollywood slumbers at the end of the Hollywood and Cahuenga Valley street railroad, out in the area where the north-south and east-west lines of the land survey begin to efface the diagonal landholding pattern of the earlier Spanish area.

The contrast between the maps of 1893 and 1966 (the area of the 1966 map is shown in outline on the 1893 map) is most striking. Now the grid has spread over virtually the whole area, with only the hills, the country club, and the expressway offering some resistance to the ubiquitous rectangularity. Still, even in the city that invented the freeway, many instances will be seen of streets following the old routes. The two major roads into the Santa Monica mountains above Hollywood are good examples.

Chapter Eleven

Twentieth-Century Highway Maps

Thomas Schlereth

*E*ven before 1900, there were maps of road systems in North America, like the Spanish maps of *El Camino Real* or Anglo-American highway systems described by Christopher Colles in his *Survey of the Roads of the United States of America: 1789* (edited by Walter W. Ristow, Cambridge: 1961). But for highway maps proper we have to await the twentieth century, with its explosive expansion of the numbers of people who needed to be able to navigate themselves regularly into unknown parts of the country. This huge expansion had its cartographic counterpart in an enormous growth of maps to help these new travelers. But these maps have remained ill-known to historians, partly because as sources they are disparate and scattered.

The immediate precursors of the twentieth-century highway maps were the bicycle maps of the late nineteenth century. Mass-production of bicycles for the U.S. market began in 1877, and in 1880 the League of American Wheelmen (LAW) was formed. This organization issued both sheet maps and road map handbooks for much of the eastern United States. Other cycling organizations and local wheel clubs also produced bicycle maps that provide the historian with a graphic depiction of the pre-automotive landscape of the late nineteenth century (Plate 11.1).

In the United States, the highways had been largely eclipsed by the expansion of the railroads since the 1850s, and the bicycle maps, with their emphasis upon route and network, began to reverse this trend, and to revive public interest in highways. They thus anticipate the cartographic format of the universally known road maps of the twentieth century, and as W. W. Ristow has noted, it was these cyclists' maps that early motorists at first used. However, specialized auto maps came very soon after the introduction of gasoline-propelled motor vehicles. The first successful machine of this kind was made in the United States in 1893, and in 1895 the Chicago *Times-Herald* published the first auto map, to trace the course of an auto race sponsored by that newspaper.

The descriptive auto route guide map was the most popular form of highway guide in the United States up to 1917. These guides contained lengthy written evaluations of specific routes, indicating road conditions, points of interest along the routes, and highly detailed descriptions (with occasional photographs) of intersections where changes of direction were necessary. In each guide, two descriptions of a route (coming and going) were included, providing the historian with a paragraph-by-paragraph summary of local and regional road environments. Here the orientation sketch maps (Plate 11.2) included in the elaborate volumes (most contained several hundred pages, some as many as several thousand) serve principally as supplements to the books' verbal data (in fact the first American auto guidebook issued in 1900 contained no cartography at all).

Social historians will be interested in the wide range of information found in these automotive Baedekers: current road ordinances, court opinions on the legal issues involved in auto-horse traffic confrontations, city driving laws, and license regulations. One also finds numerous advertisements for automobile accessories (especially for odometers), hotels, and resorts. Since early automobile associations and various companies involved in the manufacture of automotive equipment (for example, Goodrich and White Motor) were major producers and distributors of these guides, the books also give us some insight into these new aspects of North American economic history.

A specialized form of this type of cartography was the photo-auto map, which, although only briefly marketed (about 1906–1910), was one of the most original methods of early road mapping (Plate 11.3). Chiefly the innovation of the Rand McNally Company (Andrew McNally II and his bride photographed the firm's first series of the Chicago to Milwaukee route on their honeymoon), the photo-auto map combined the three methods of highway guidance (verbal, pictorial, and cartographic) contending for the motorist's attention in the early twentieth century.

The photo guides are a vernacular architecture historian's delight because their abundant "photographs at every turn" (two to a page with a center section left blank for the motorist to record personal observations of the landscape and touring experiences) help docu-

ment, at precise times and specific places, the built environment of urban and rural North America. Houses, fence types, farmsteads, barn types, vegetation, signage, and often intersection-by-intersection detail of urban spaces are provided. Captions provide commentaries on parks, cemeteries, scenic sites, garages, overnight accommodation, and other service facilities. A photo guide map that has been annotated as a motoring diary can be extremely useful to the historian for certain topics in social history.

Each photo-auto map volume contained crude sketch maps drawn as strip maps; they were designed to link the book's photographs and to give the motorist an indication of direction of travel as well as a computation of distance between turning points. Although not drawn to scale, the maps showed the location of all railroad crossings, bridges, school houses, and other key landmarks. The photo maps also include written descriptions of the routes traversed; these accounts, along with the photographs and the sketch maps in the guide, had to be researched for two routes (out and back). Thus the photo map, like the verbal auto guide, provides the historian both narrative and graphic data with which to interpret a landscape route.

Only major traffic routes east of the Mississippi and north of the Ohio rivers were mapped as photo-auto guides by Rand McNally (approximately twenty-five books) and a few other firms (White Motor Company, Goodrich Motor Company). High printing costs, continually changing landmarks, and bulkiness (the 1909 Chicago to New York map guide was over 200 pages) eventually eliminated this road map type from competition with sheet maps. However, its simple strip map reappeared, in more sophisticated form, in the "Triptik" road map format prepared by the American Automobile Association since the 1950s.

In 1907 the first gasoline station chain, American Gasoline Company, began in St. Louis, and the following year Ford's model T began mass production. From being a rare and somewhat adventurous novelty, the automobile was becoming a conventional means of transport for work and recreation, and the maps naturally reflected this. In 1911 the National Highways Association was formed, and the following year the Lincoln Highways Association began marking the roads with distinctive signs. This vogue for the "blazed trail," with brightly colored identification placards, bands, or blazes, gave rise to the blazed trail map (Plate 11.5). Over thirty principal highways and 150 local routes were thus blazed up to 1930, and the names given various localities on the ground and on the map offer the place-name student an interesting, if brief, glimpse into local and regional toponymy.

On our example, Texas state highways are shown with their numbered markings, recalling that it was the states (Michigan first in 1920; twenty-four other states by 1925) who initiated numerical road identification systems several years before the system begun by the federal government in 1925. A useful resource in the historical study of early twentieth-century tourism, the blazed trail map has a modern analogy in the state highway maps that many states use to promote tourism on routes with such names as the Lincoln Heritage Trail (Illinois), or the Spanish Mission Trail (California), or the Trail of Independence (Texas).

From that first station chain founded in St. Louis in 1907, various chains had been spreading rapidly over the country. In 1913 the Gulf Oil Company began station franchising, and the following year William B. Akin, head of a Pittsburgh advertising agency, persuaded that company to prepare a free road map to be mailed to ten thousand registered car owners in the Pittsburgh area. So that most ubiquitous of all North American maps, the give-away highway map, was born, and thrived until comparatively recently.

A sequence of these maps can provide us with a wealth of geographical, historic, and cartographic information. Issued and corrected annually, these maps include thousands of changes from each year's previous issues. One can chronicle highway expansion, route improvement, and road abandonment as well as monitor the evolution of such important urban freeway forms as the Los Angeles freeway system depicted on Plate 11.6. Reviewing annual highway maps also provides us with clues as to changes in settlement patterns, network expansion, population shifts, land use, and recreational facility needs.

Most of the cartographic symbols found on contemporary U.S. maps already appear on Plate 11.6, a map of 1936. When compared with maps of this vintage, modern highway maps are seen to resort to increasingly abstracted symbolization for identifying roads, and to increasingly complex classifications for describing their state (a dozen or more classifications now being possible). In addition to some description of topographic features, such maps indicate highway construction and alternate routes. Sometimes these sheet maps have been combined into road atlases, of which the first was Rand McNally's national highway atlas, first marketed in 1924 as the *Rand McNally Auto Chum* (see Plate 11.4).

One final type of road map is the county transportation map, of which the first was prepared by the state of Michigan in 1919. These large-scale maps are among the most detailed types of highway cartography generated by government agencies, only the USGS maps (see Chapter 10) normally showing more detail. They are indispensable to any historian of local roadside life as they pay detailed attention to specialized material culture such as drive-ins, trailer courts, motels or tourist courts, highway garages, and wayside parks. Settlement configurations and economic activities of unincorporated towns and villages, as well as the design formats of interstate highway exchanges, are often mapped in special inserts (Plate 11.7).

The historical value of these county maps is primarily due to their large scale, as they are prepared in order to be used by county highway road departments. In addition to their wealth of cultural data, they thus provide the most up-to-date road information on all primary and secondary roads in the nation. On few other maps can one find the spatial configurations such as migrant worker housing, Grange hall location, or landfills. The Maryland state highway administration recently (1980) began to include multicolor relief information on their maps, making them even more valuable as topographical and historical resources. In many states, the information from these specialized and large-scale maps has been combined onto sheet maps of the state at roughly the same size as the old oil company maps. These state maps, generally distributed by the state tourism authorities (Plate 11.8) have thus in a sense replaced the oil company maps. They resemble them fairly closely, but generally have slightly more adequate indications both of hydrography and of cultural features.

The scholarly pursuit of the abundant yet elusive highway map is a research enterprise with both cartographic and historical significance. Road maps have some very distinct characteristics. First, they are designed for regular use primarily in guiding motorists, and not for casual reference or direct research purposes. So the road map must embody features and information in such a way that it can be read very quickly by a motorist traveling without recourse to any other printed sources of information. Second, design features must be constructed for the nonprofessional user, necessitating the simplification of many symbols and the addition of many features to help the average motorist. Third, because these maps are directed to the general public, they must be both attractive and modern.

Comparing road maps to other types of map, one can readily see that they use many of the most modern and sophisticated techniques: half tone, color, differentiation between the size of urban settlements through boldness of print rather than size of it, and specialized symbols indicating types and conditions of highways. They are thus useful to the historian not only as sources of information concerning the land they portray, but also as precious indications of what mapmakers thought their users needed—an insight into the presumed cartographic capacities of the average citizen in the late twentieth century. They are also repositories of popular art motifs, as Color Plate 6 shows. Before the age of mass air travel, these maps contrived to make a romance out of voyages in this country, and this sense of adventure can still be felt from some of the surviving maps.

SOURCES AND STUDIES FOR THE HIGHWAY MAPS

HISTORIES OF THE HIGHWAY IN THE UNITED STATES

Hoy, Suellen M., and Michael C. Robinson, eds. *Public Works History in the United States*. Nashville: 1981. Chapter 8 contains a bibliography on roads, streets, and highways.

Labatut, Jean, and Wheaton J. Lane, eds. *Highways in Our National Life, A Symposium*. Princeton: 1950. This volume gives a general view of the theme.

Mumford, Lewis. *The Highway and the City*. New York: 1963. This book is useful as a strongly interpretive account of the relationship.

Pawlett, Nathaniel M. *A Brief History of the Roads of Virginia, 1607–1840*. Charlottesville: 1977.

Pierson, George W. *The Moving American*. New York: 1973.

Quaife, Milo. *Chicago Highways Old and New: From Indian Trail to Motor Road*. Chicago: 1923. An early history that has still not been superseded.

Rae, John B. *The Road and the Car in American Life*. Cambridge, Mass.: 1971.

Riesenberg, Felix, Jr. *The Golden Road*. New York: 1962. The concluding chapters contain excellent research on twentieth-century highway history.

Rose, Albert C. *Historic American Roads: From Frontier Trails to Superhighways.* New York: 1976.

Schlereth, Thomas J. *U.S. 40: A Roadscape of the American Experience.* Bloomington/Indianapolis: 1985. Part III of this work offers a historiographical overview of the entire topic.

Stewart, George. *U.S. 40: Cross Section of the United States of America.* First published Boston: 1953; reprinted Westport, Conn.: 1973.

U.S. Department of Transportation—Federal Highway Administration. *America's Highways, 1776–1976: A History of Federal Aid Program.* Washington, D.C.: 1976.

Vale, Thomas, and Geraldine Vale. *U.S. 40 Today.* Madison: 1984.

WORK ON DIFFERENT TYPES OF HIGHWAY MAPS

Bay, Helmuth. "The Beginning of Modern Road Maps in the United States." *Surveying and Mapping* XII (1952), pp. 46–48.

California State Automobile Association. "The History of Road Maps." *Motorland* LXXXI (1960), pp. 6–8.

Ladd, Richard. *Maps Showing Explorers' Routes, Trails and Early Roads in the United States.* Washington, D.C.: 1967.

MacDonald, Thomas H. "Map Work of the Public Roads Administration." *The Military Engineer* XXXII (1940), pp. 37–38.

McKenzie, Roderick Clayford. "The Development of Automobile Road Guides in the United States." Master's thesis, University of California at Los Angeles, 1963.

Pawlett, Nathaniel M. *A Guide to the Preparation of County Road Histories.* Charlottesville: 1979. A treatment of government-highway road mapping at the local level.

Rand McNally. *Guide to the Blazed Trails.* Chicago: 1917.

Ristow, Walter W. "American Road Maps and Guides." *Scientific Monthly* LXII (1946), pp. 397–406.

———. "A Half Century of Oil-Company Road Maps." *Surveying and Mapping* XXIV (1964), pp. 617–37.

———. "Maps for Extra Motoring Pleasure." *Fine Cars,* July 1954, pp. 10–12.

———. "Thank the Wheelmen for your Road Map." *Frontiers* XI (1946), pp. 56–59.

Schultz, Gwen M. "New Developments in American Road Maps." *The Professional Geographer* XV (1963), pp. 13–14.

Wise, Donald A. *Sources of Official State Maps.* Washington, D.C.: 1975.

REPOSITORIES AND ARCHIVES

Most extant highway maps can be divided into two broad categories: bound volumes and sheet maps. Bound volumes in the form of road guides, triptiks, or road atlases are more likely to have survived, and various repositories, public and private, have at least fragmentary holdings. Complete collections of this material are, however, rare, for most repositories, especially libraries, regard even bound volumes of highway maps as ephemera, and in any case the sheer number of bound road guides issued during this century defies the possibility of complete collection.

Sheet maps are even less likely to have survived in complete collections. Their very nature is a hindrance to curators, for a sheet of paper, folded and refolded many times, rapidly disintegrates. Since highway sheet maps are designed to be handled in this way, many do not survive their original owners. In certain archives where adequate conservation measures have not been taken, the researcher can also expect to encounter highway sheet maps so badly worn as to make them almost unusable.

Despite these problems, there are significant caches of twentieth-century highway maps in various institutions and organizations. A largely unknown amount also survive in private hands in the form of touring scrapbooks, vacation memorabilia, office files, and individual collections of private map aficionados interested in the North American car culture or the history of the North American road and roadside. Most accessible to the general historian are highway maps housed in institutional contexts that permit research use on a regular basis. Such repositories can be characterized as governmental, corporate, library, and private association.

The chief governmental repository is the Geography and Map Division of the Library of Congress. This huge library has a very good collection of North American highway maps, and there are other federal agencies that hold this type of material.

These include the Topographic Center of the Defense Mapping Agency (see the DMA's *Price List of Maps and Charts for Sale*); the U.S. Army Corps of Engineers, the Tennessee Valley Authority, the Department of Agriculture, the Bureau of Land Management, the U.S. Geological Survey, the Bureau of the Census, and the Department of Transportation. Donald A. Wise's booklet, *United States Official Mapping Agencies* (Washington, D.C.: 1977) and Morris M. Thompson's Chapter 8, "Maps from Other Agencies," in *Maps for America: Cartographic Products of the U.S. Geological Survey and Others* (Washington, D.C.: 1979) are two guides to these varied repositories.

The Federal Highway Administration (Office of Public Affairs, Room 4208, 400 7th Street, S.W., Washington, D.C. 20590) of the Department of Transportation contains a wide range of highway cartography. For example, sequences of county highway maps showing all principal primary and secondary roads in the United States can be obtained from this agency; also available are highway maps dealing with national road networks from the beginning of the federal highway numbering system in the 1920s to the development of the Interstate System. Needless to say, many other government agencies also prepare highway maps, and information about these can often be obtained from the state highway (or transportation) department.

Among the corporate repositories, the archives of the private oil companies are a major institutional resource. Such firms as Mobil, Sunoco, Texaco, and Amoco all issued free road maps, which they may have on file. Nor should the historian neglect the commercial firms of automobile highway mapping: Rand McNally (Archives and Library, Skokie, Illinois); General Drafting Company (Archives, Convent Station, New Jersey), and H. M. Gousha (Archives, San Jose, California). For the names and addresses of other commercial houses that have published North American highway maps, see Donald Wise's booklet, *United States Private and Commercial Map Publishers*. (Washington, D.C.: 1975).

Among the libraries, research institutions such as the University of Wisconsin at Milwaukee (home of the Map Collection of the American Geographical Society), the Western Reserve Historical Society (Cleveland, Ohio), the University of Illinois, the Museum of the National Road (Zanesville, Ohio), the Division of Transportation, National Museum of American History of the Smithsonian Institution, and the Robert Tauthill Library (the Edison Institute, Dearborn, Michigan) have useful collections of road maps.

National, regional, and local automotive associations constitute a final resource for highway maps. National associations such as the American Automobile Association (8111 Gatehouse Road, Falls Church, Virginia 22042) and the National Geographic Society (17th and M Streets, N.W., Washington, D.C. 20036) produce and distribute a variety of maps and related materials. Detailed information on the kinds of cartographic products or services available may be obtained directly from each firm or national association.

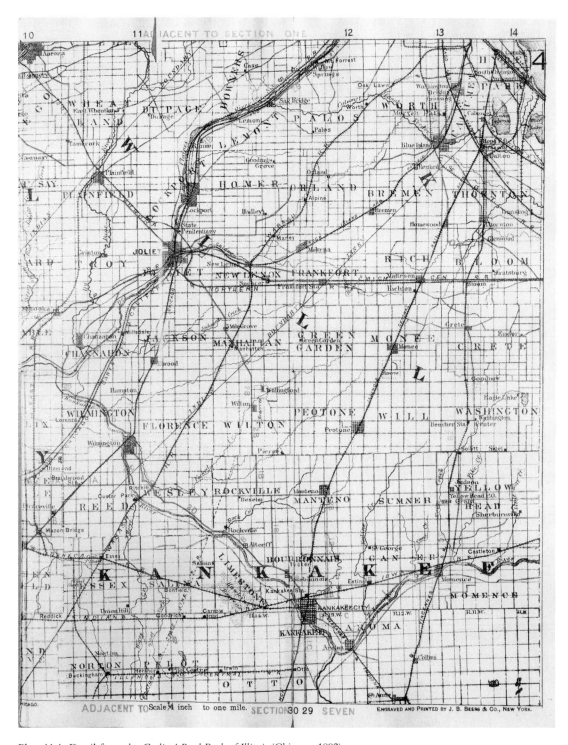

Plate 11.1. Detail from the *Cyclists' Road Book of Illinois* (Chicago, 1892).

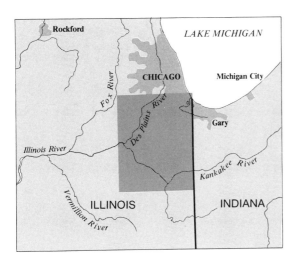

Plate 11.1a. Modern map to show location of Plate 11.1.

Bicycling through Illinois in the 1880s

This is an excerpt from the regional route map published by the League of American Wheelmen (LAW) in 1882. In this part of Illinois, the usual evaluations of terrain—rolling, hilly, and so forth—were hardly necessary, but the map contains much of the other information found on maps of this kind including a rudimentary road classification, computations of distance between settlements, and local landmarks. Some routes are marked "Century Rides" since an objective of many cyclists was to do a one hundred–mile round-trip or one-way route whenever possible in a day's outing.

Cultural data such as rural railroad stations or stops, post offices, schools, and unincorporated villages are usually noted; commuter and interurban rail lines are also depicted, since cyclists often used them in order to reach special touring roads, or returned home on them after a strenuous day on their wheels. This interrelationship of rail and road networks, found on most bicycle maps, can also be used by the student of urban expansion as another description of the processes of late nineteenth-century suburbanization. A useful historical comparison, easily adaptable as a classroom exercise, is to contrast early LAW and other local bicycle maps with modern examples such as the work by Gershon Weltman and Elisha Dubin, *Bicycle Touring in Los Angeles* (Los Angeles, 1979) maps, or the cartography prepared by the American Bicycle Association.

Plate 11.2. Detail from the general index map in the *Official Automobile Blue Book 1917* (New York, 1917).

Traversing the Wilds of Northern Michigan, 1917

This is an example of an auto route guide, a genre that was very popular up to about 1920. The cartographic element is relatively minor, consisting essentially of rather basic maps with which to make sense of the abundant text. This is full of phrases that seem to hint at the relative inadequacy of the vehicles of the time, given the condition of the roads. Route 993 on our map, for instance ("Gaylord to Bay City"), is described as "Easy sand across stumpland to Prescott and gravel and stone balance of way." We seem to see the anxious motorist, waiting for sand that will not be easy, and navigating among the stumps of the cut-over forest.

Route 893 ("Benzonia to Traverse City") brings out the enthusiasm of the commentator: "Around the Horn. A splendid scenic trip thru the lake country of Leelanu County. Gravel and sandy gravel, not by any means bad. A wonderful trip thru a wild unfrequented country among beautiful lakes and attractive resorts." This was the country in which the young Ernest Hemingway was growing up, on his vacations from Oak Park in Chicago, and we still catch some of the romance of that period, when a trip to upper Michigan was an adventure, and a motoring journey across the whole country virtually unthinkable.

No. 38

Turn RIGHT, East. Small lake on left, half a block distant, large hole on left after turning. Go two blocks then turn left and straight ahead to Main Street. Go straight through town, passing Court House. Next photo three and two-tenths miles. For Automobile Maintenance Garage turn to right one block beyond Court House and go one half block. This garage is well equipped and is open day and night.

Memo ....

No. 39

Take LEFT hand road. Follow poles with most wires. Next photo four and five-tenths miles.

No. 40

STRAIGHT AHEAD, North-East. Next photo South Bend twenty and five-tenths miles. Before coming to South Bend New Carlisle is passed, five and two-tenths miles. Follow the wires.

Memo ....

No. 41

SOUTH BEND

Turn RIGHT, South. Turn from La Salle Street into Main Street, go two blocks to Hotel Oliver.

Plate 11.3. *Photo-Auto Map: Chicago, Illinois to South Bend, Indiana/South Bend, Indiana to Chicago, Illinois* (Chicago, 1907).

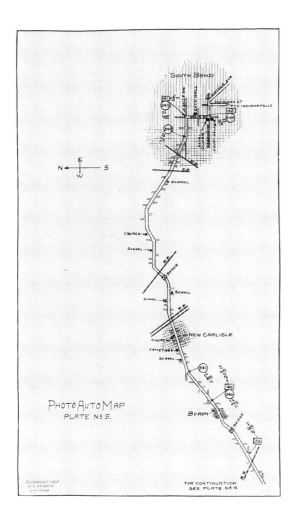

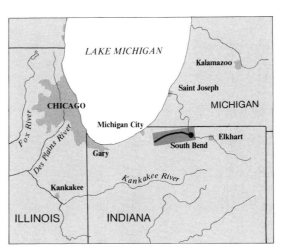

Plate 11.3a. Modern map to show location of route shown on Plate 11.3.

Detailed Instructions for Navigating from Chicago to South Bend, 1907

Like the verbal auto guide shown on the previous plate, the photo-auto map was an ingenious way of trying to combine verbal and pictorial evidence. Plate 11.3 shows a section of the route from Chicago to South Bend, Indiana, with specific instructions at each intersection. One has to imagine that not many journeys as yet took place at night, when such instructions would not have been very useful.

Left above is a section of the type of strip map that usually accompanied these photo maps. It shows the last part of the road into South Bend, from the same period as Plate 11.3, and is typical of the careful description needed before roads had been numbered. We have to remember that motorists at this time would probably be driving at about twenty miles an hour, looking out for those "bad patches," and that decisions consequently did not have to be made in any great haste.

Plate 11.4. Detail of Florida from the *Auto Road Atlas of U.S.*, Copyright 1926 by Rand McNally, R.L. 89-S-120 (Chicago, 1926).

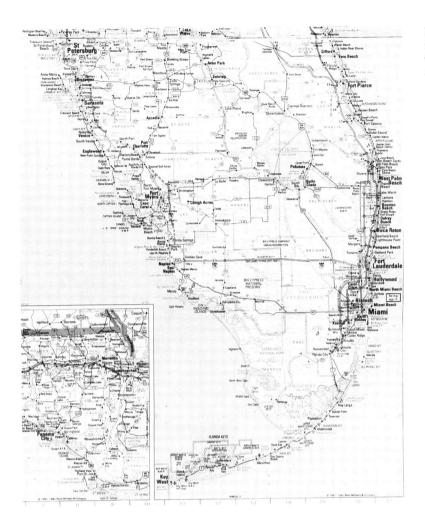

Plate 11.4a. Detail of Florida
from the *Road Atlas*.
Copyright © 1981 by Rand
McNally, R. L. 89-S-120.

Southern Florida in the Early Automobile Age

Road maps, frequently consulted, are by definition vulnerable, even in atlas form, and it now seems possible to locate only one copy of Rand McNally's first road atlas, the *Rand McNally Auto Chum* of 1924. That copy is so battered that we have chosen here to reproduce a plate from Rand McNally's *Auto Route Atlas of the United States,* published in 1926 and reproduced in facsimile in 1974.

The image of southern Florida in 1924 has changed less, in many ways, than one might have thought, to judge from the corresponding plate in Rand McNally's road atlas of 1982 (Plate 11.4a). There are, it is true, many more roads and more settlements, as well as many interesting changes of name, but the interior of the state appears to be more or less untouched by major roads. That does not mean, of course, that it is untouched by the development on all sides of it, but we would need a different type of map to track those environmental changes.

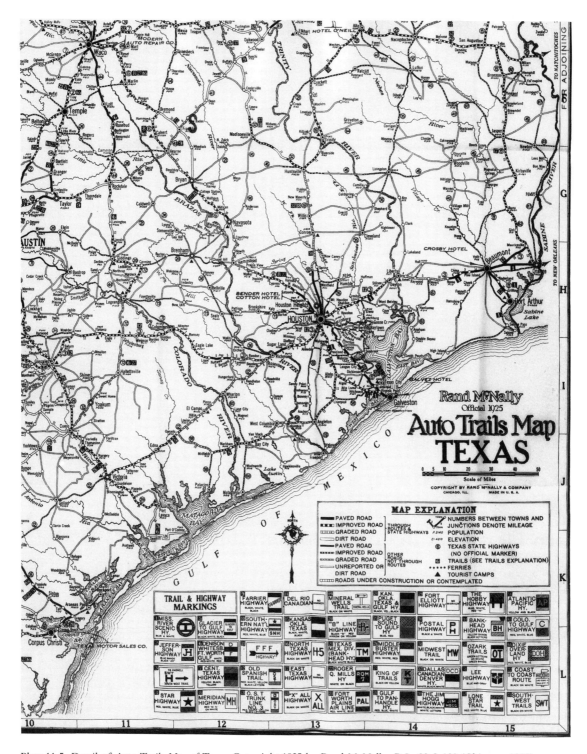

Plate 11.5. Detail of *Auto Trails Map of Texas.* Copyright 1925 by Rand McNally, R.L. 89-S-120 (Chicago, 1925).

Plate 11.5a. Modern map to show area of Plate 11.5.

Following the Blazed Trails in Texas, 1925

This map, published by Rand McNally a year before the federal highway numbering system was inaugurated, shows blazed trail mapping at its zenith. There is very detailed information on the state of the roads, which indeed do not seem very numerous, and some indication of population, as well as of the location of tourist camps, garages, and hotels. It is easy to see from the key why this is called an auto trails map, for the number and variety of the trail and highway markings were beginning to become bewildering. The inauguration of a federal numbering system was almost inevitable, if only to avoid the sort of confusion that this proliferation would have caused.

The grading of the roads merits close analysis, in conjunction with a modern road map of the same area. In fact, it would be a good classroom project to consider whether the main roads shown here go back to Spanish times, and in response to what needs (tourism? agriculture? early oil drillings?) the new ones were now being made.

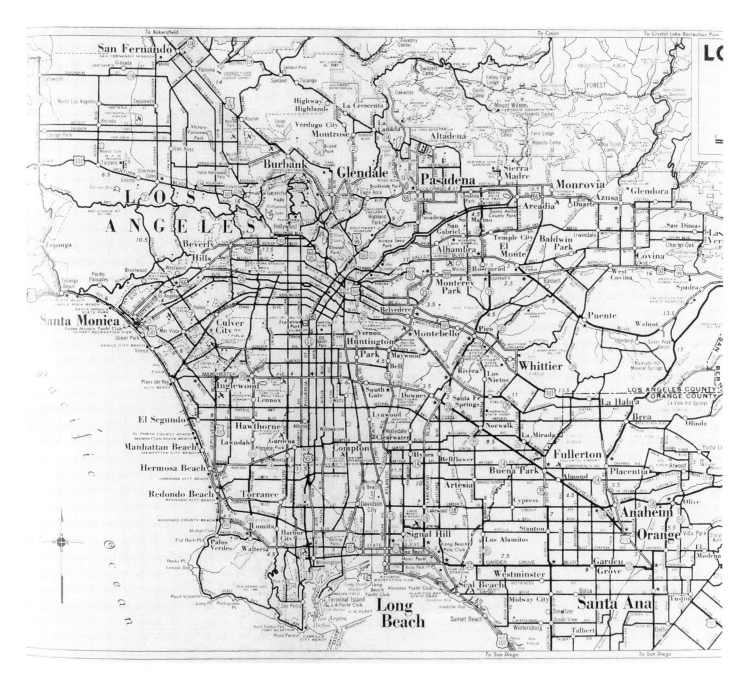

Plate 11.6. Detail of the "Highway Map of Los Angeles and Vicinity," from the *State Farm Auto Insurance Highway Atlas* (Chicago, 1936).

Plate 11.6a. Map of "Los Angeles and Vicinity," from *California, Nevada* (road maps issued by the Exxon Oil Company in 1975). Map courtesy of General Drafting Company, Inc.

Toward an Increasingly Abstract Type of Road Map, 1936

Although it was drawn over fifty years ago, most North Americans would immediately recognize and comprehend this map type: the complimentary oil company map. This rather austere and increasingly symbolic type of map eventually replaced the older guidebook maps, with their running commentaries and the relatively personal blazed trail maps. Here we have an arrangement of streets, greatly exaggerated in their girth, identified by numbers and linking townships identified as to size by the grading of their typography.

There is very little feel for the topography of the country, but much evidence of the way things are going, in the proliferation of golf courses and country clubs, as well as in the charting of access to private and municipal airports. Rand McNally first prepared this map as a sheet map for free distribution by the Mobil Oil Company, and then used a similar version in the edition of the road atlas that the firm printed for the State Farm Insurance Company. This was indeed a feature of the oil company give-away maps, that they tended eventually to be collected as well into road atlases, which could be left in an automobile and used for reference wherever the driver might be in the nation.

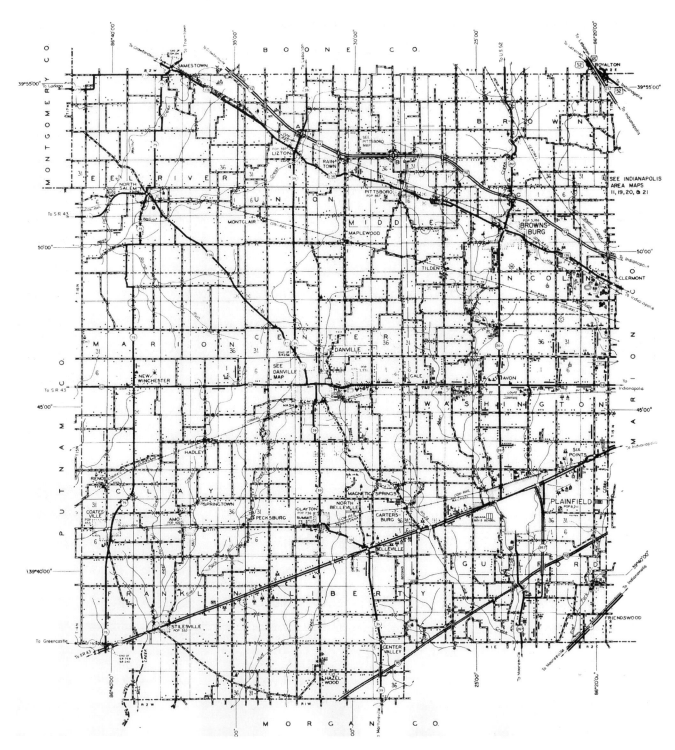

Plate 11.7. *General Highway and Transportation Map, Hendricks County, Indiana* (Indiana State Highway Commission, 1976).

Close Detail of Hendricks County, Indiana, 1976

The various highway administrations within each state have been producing for the past seventy years maps like this one, with close detail on the roads and an indication of each building. They also have insets with schematic plans even of very small towns. If studied in the order of issue, county transportation maps like these can demonstrate agricultural, commercial, industrial, and residential change at a microlevel. They were eventually collected at the Federal Highway Administration in the U.S. Department of Transportation, and there provide the most recent information on all the roads in the nation.

This particular map shows part of Indiana within the area in which the township and range system profoundly marked the country, so that the question in interrogating a map like this is, what has given rise to the occasional diagonal cultural feature? Many states have this type of map bound together in a spiral binder, and coverage of this type is the easiest way to carry into the field a detailed summary of all the counties in any given state. Of course, for really close work in the field, it will be better to use the appropriate USGS map (see Chapter 10).

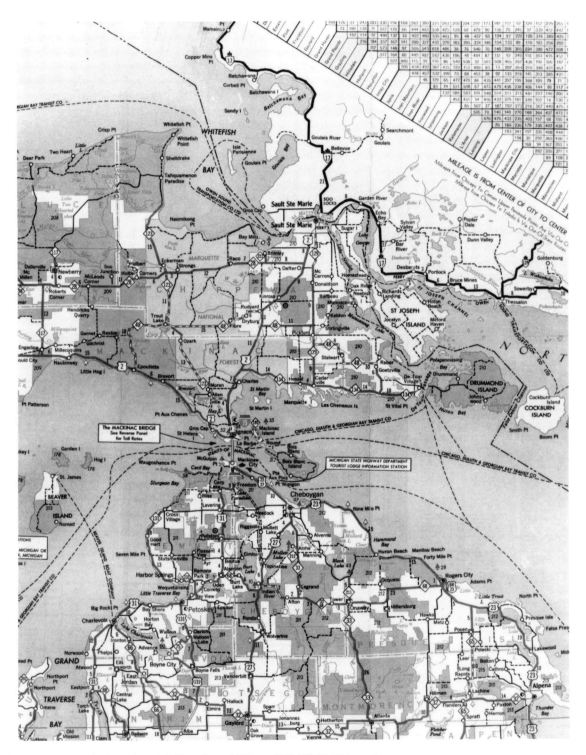

Plate 11.8. Detail of Northern Michigan from *Michigan: 1960 Official Highway Map* (Michigan State Highway Department, Lansing, 1960).

Present-Day Image of Part of Northern Michigan

With the demise of the give-away oil company maps, many states provide in their tourism offices something like this official highway map from Michigan. It is issued by the Michigan State Highway Department, and was no doubt generated from large-scale maps like the one discussed as Plate 11.7. These maps are very similar to the old oil company maps, but they do tend to show slightly more in the way of hydrography and vegetation. Of course, they remain mute on questions of topography, for which the USGS map at some scale or other is essential.

It is interesting to compare this plate with Plate 11.2, from the early guidebook. Whereas the former was intended to be schematic, and showed only a very few roads (though there *were,* indeed, fewer roads here at that time), this newer plate shows a map that tries to be complete in itself. It is true that you can use this type of map to drive to virtually any locality of any size, but it seems a pity that we have lost the concept of having a gazetteer to accompany the map. For now, in order to find out about the many towns encountered along the way, we should have to have recourse to something like a state guidebook or WPA history, and very few travelers take the trouble to do that.

CHAPTER TWELVE

Aerial Imagery

David Buisseret and Christopher Baruth

It has long been realized that some places are best viewed from the air. As we have seen, the bird's-eye view became a standard device for portraying cities, and there were occasional attempts to show countrysides in the same way. Among the greatest artists, for instance, both Dürer and Leonardo depicted landscapes from a high and sometimes imaginary angle. A relatively late but very remarkable example of this type of depiction is George Catlin's watercolor "View of Niagara" (Plate 12.1). This extraordinary work, drawn using a model, most curiously prefigures the aerial photographs that began to be made from the middle of the nineteenth century onward. It is not a truly vertical image, but to read many parts of it we need to perform the same type of mental adjustment that we have to carry out in order to relate aerial photographs to topographical reality. Of course, what took Catlin much time and originality eventually came to be performed at the click of a shutter.

From the earliest origins in nineteenth-century France, aerial views have become varied and numerous, and in the past thirty years they have come to include views from space. However, all this material may be divided into three main categories, according to:

1. The height of the imaging vehicle.
2. The angle of vision.
3. The type of image produced.

The highest flying vehicles are of course spacecraft, whether manned or unmanned, and these customarily collect their images between 150 and 500 miles away from their targets on the earth. Lower down, we have high-flying aircraft, which carry out missions somewhere between 5,000 and 60,000 feet. Lower still, and down to ground level, we have other types of aircraft, both fixed-wing and helicopters and balloons, which can offer greater details.

The angle of view may be either vertical or oblique. Virtually all images made above 5,000 feet are vertical, but below that height either type of imagery is common. As we shall see, vertical images are most useful for such purposes as accurate cartographic plotting, while oblique photography is often used for detecting or illuminating specific sites.

There are four main types of image. The first and oldest is the simple black-and-white photograph, of the type taken throughout the nineteenth century. Later on it became possible to produce color photographs, and these too were eventually used in aerial photography. A third type of image is produced by the

Plate 12.1. "Topography of Niagara," watercolor by George Catlin (private collection).

David Buisseret and Christopher Baruth

use of multispectral scanners, which not only detect visible light but also distinguish between the heat emissions of different types of features, and can collect reflected radiation in the near infrared; much space imagery is of these types. The fourth category of images comprises those produced by more or less exotic sensors such as radar; use of these images has so far been largely confined to military purposes.

Looking at an aerial image, then, we need to bear the three criteria in mind. For instance, plate 12.4, showing Green Bay in Wisconsin, is derived from a vertical space view, using the Landsat Thematic Mapper (a type of multispectral scanner). Plate 12.7, on the other hand, was obtained from a low-level black-and-white photograph. These different types of images have widely varying uses, as we shall now see.

In general terms, aerial images offer three types of advantage over conventional ground-level views.

1. Aerial images sometimes permit us to detect phenomena invisible on the ground. This was the classic use of low-level black-and-white photographs, whether vertical or oblique, in the study of Roman sites in western Europe. Using such signs as differential crop growth, variable snow cover, and so forth, the pioneers like Agache, Crawford, and Scollar were able to identify many sites that never could have been picked up merely through ground inspection.

2. Aerial images give instant, accurate information about transient phenomena. For cartographers and geographers it has been invaluable to have instant images of sedimentation, ocean currents, or pollution, phenomena that would be almost impossible to plot accurately using any other means. For the historian, too, such transient phenomena can be very revealing, as for instance in the differential crop growth that enables us to distinguish international boundaries, or to detect Indian settlements within the United States (Plate 12.5).

3. Aerial images often offer us a new view, either because of their all-around coverage, or because of their novel angle, or because of their great extent. The all-around coverage is most striking in the case of buildings, which can often be set into their context in a way impossible with a ground view (Plate 12.8). The novel aerial angle sometimes allows us to identify field patterns, or ground plans, occasionally for sites inaccessible from the ground. The great extent covered by many aerial photographs allows us to track over very long distances such wide-ranging features as Roman roads, or midwestern landholding boundaries.

These general uses may be translated into some specifically North American applications. First of all, wide-ranging images allow us to gain an unrivaled grasp of the lie of the land. With multispectral scanning, we can identify the gross geography of an area and its main hydrographic features, and then go on to examine the nature of its vegetation. Plate 12.2 offers an example of this for a relatively restricted area, in which the main hydrographic features are particularly inconspicuous from the ground.

Having identified the natural features, we can then analyze the network of communications that previous settlers have often laid upon the surface of the land. Plate 12.2 is again a good example of this; we see the spokelike pattern of roads radiating from the city of Chicago, and using a sequence of historic maps, we can trace them back to the plank roads of the middle of the nineteenth century, and often back before that to the trails of the Indians. On the same plate we can also trace not only the natural waterways that made Chicago the natural center for the Midwest, but also the canals which confirmed that vocation: the Illinois and Michigan Canal, the Cal-Sag Channel, and so forth.

Intimately connected with the communications network are the patterns of landholding. In North America, the regional varieties of these patterns tell us much about early settlement. In the East, boundary arrangements were often informal and apparently disorganized, as they were in parts of the West. But much of the Midwest is covered by the township and range grid described in Chapter 4 and exemplified in Plate 12.4, where indeed it is broken by the French system of long lots, evident in many sections of the arc from Québec to Chicago and down to New Orleans. In the West, the township and range system eventually came up against the existing Spanish land grants, which adhered to a different pattern (Plate 12.3). One way and another, a survey of the pattern of landholding, made relatively easy by the use of aerial images, tells us much about the early conditions of settlement.

In order to detect Indian survivals, we generally have to use images captured closer to the ground. Many of the great Indian earthworks have been destroyed, and others, in spite of their great size, do not show up well from the air. But there are some well-preserved geometrical figures such as the ones on Plate 12.7 that can best be appreciated using low-level vertical aerial views. No doubt many more such geometrical figures remain to be discovered.

Sometimes the early phase of European occupation can also be well documented from the air. Plate 12.10 shows a fortification con-structed during the Civil War at Fort Union in New Mexico; it looks quite incoherent from the ground, but from the air is seen to be modeled on the lines of the classic European star forts. In the background, the desert is crisscrossed by many sections of the Santa Fe Trail. It is odd that no attempt seems to have been made to track the Santa Fe and Oregon trails using photographs of this type. Plate 12.8 illustrates another area of European settlement, showing a plantation house on the Mississippi River. Because of the height of the levee, it is impossible from the ground to understand the relationship between the house and the great river, but this emerges at once from the aerial view.

Sometimes a low-level oblique view can be revealing about quite recent works. Plate 12.11 shows the landfill at Chicago in 1930, when preparations were being made for the 1933 World's Fair. It is a fine example of the way in which a transient phenomenon can be captured at a certain stage, giving us a strong feel for the scale of an enterprise like this.

Aerial photographs, then, have much to offer the historian, and it is certain that much work remains to be done in the United States to exploit their full potential. In the following pages, we try to offer examples of how they can be used, with suggestions for further investigation.

SOURCES FOR AERIAL IMAGERY

There are very few publications that deal with the whole range of aerial images, from space views down to low-level photographs. One remarkable exception is *Cartography and Remote Sensing Imagery* (in the Washington, D.C. area), edited by Ralph Ehrenberg (Washington, D.C.: 1987), which gives an extraordinary summary of the great wealth available in the nation's capital. Generally, though, in looking for these images we have to search the publications under at least three different categories:

1. Space images.
2. High-level vertical photographs.
3. Low-level oblique photographs.

The number of publications devoted to these three categories is very uneven, and here we shall simply try to pick out some of the more useful finding aids and some of the best collections of imagery.

SPACE IMAGES

Almost all images used in North America have come from the LANDSAT system, though in the past year or two it has also been possible to obtain material from the (French) SPOT Image Corporation. Information about LANDSAT is most easily available from the various National Cartographic Information Center (NCIC) offices, which are part of the National Mapping Program. The main NCIC office is at the U.S. Geological Survey, 507 National Center, Reston, Virginia 22092; material may also be ordered directly from Eosat (formerly NOAA Landsat) Customer Services, Mundt Federal Building, Sioux Falls, South Dakota 57198.

The NCIC publishes a number of useful leaflets on LANDSAT and adjacent subjects. Some of the most interesting are:

How to Order LANDSAT Images.

Manned Spacecraft Photographs and Major Metropolitan Area Photographs and Images.

A Selected Bibliography on Maps, Mapping and Remote Sensing.

Understanding Color Infra-red Photographs and False-color Composites.

The SPOT Image Corporation will also provide imagery, sometimes of an exceptionally high level of color and resolution. The Corporation publishes a quarterly newsletter called *Spotlight,* and may be reached at 1897 Preston White Drive, Reston, Virginia 22091.

During the past twenty years there have been many attempts to present satellite-derived imagery in an attractive and interesting published form. Many of the resulting books have been primarily geared towards natural scientists, but some have great interest for historians. What follows is a subjective sampling of these publications.

Avery, Thomas, and Thomas R. Lyons. *Remote Sensing: A Handbook for Archeologists and Cultural Resource Managers.* Washington, D.C.: 1977. As its subtitle indicates, this manual is primarily designed for archeologists and persons interested in preservation. But it also has sections of interest to historians, and gives a good general view of the subject.

Bodechtel, Johann, and Hans-Gunter Gierloff-Emden. *The Earth from Space.* New York: 1974. This book, originally published in German, is remarkable for the skill with which the authors, primarily interested in geology and meteorology, commentate the excellent NASA images.

Fitzgerald, Ken. *The Space-Age Photographic Atlas.* New York: 1970. A good collection of images, from LANDSAT down, of various parts of the world; the author's concern is not primarily historical development, but the images inevitably illustrate this in many cases.

National Geographic Society. *Atlas of North America; Space Age Portrait of a Continent.* Washington, D.C.: 1985. This is a varied and splendidly reproduced selection of satellite images, chosen to bring out the salient features of each region. A basic tool for historical studies of North America.

Sheffield, Charles. *Earthwatch.* London: 1981. An excellent selection of LANDSAT images, well produced, often showing phenomena of interest to the historian such as the well-defined "frontier" between Canada and the United States, or the view of Libya with clouds forming above the irrigated areas.

————. *Man on Earth.* New York: 1983. Probably the best existing set of satellite images commentated partly from the historical point of view. The reproduction is impeccable and the accompanying maps make it easy to follow the author's argument.

Short, Nicholas, et al. *Mission to Earth; LANDSAT Views the World.* Washington, D.C.: 1976. This is a useful compendium, but it is not strong on historical themes.

HIGH-LEVEL VERTICAL PHOTOGRAPHS

This type of photography has wide uses for historians and archeologists, but there are virtually no publications grouping large numbers of such photographs, no doubt because they are useful more for specific problems than for browsing. In order to find high-level vertical photographs of sites in the United States there are two helpful pamphlets, both of which may be obtained from the EROS Data Center, U.S. Geological Survey, Sioux Falls, South Dakota 57198:

How to Obtain Aerial Photographs.

The Sky's the Limit! The National High Altitude Photography Program.

This latter pamphlet, which comes with interesting examples, explains how the National High Altitude Photography Program was set up in order to save funds by bringing under one organization the various photographic programs offered by different Federal agencies. For historians, the materials made available by the Department of Agriculture, the Department of Defense, and the Department of the Interior are particularly useful. Information about all of them may be obtained from the EROS Data Center.

LOW-LEVEL OBLIQUE PHOTOGRAPHS

This is the type of photograph with which the whole range of modern aerial images began. So many libraries and archives hold more or less random collections of low-level oblique photographs that it is impossible to suggest any way of systematically examining them all except by individual visits. However, if this type of photograph is the most difficult to track down, it is also the one that has given rise to the greatest number of history-related publications. For the sake of logic, we have divided a selection of these publications into three headings: classic theory and practice, some examples from France and Britain, and finally some examples from the United States.

Classic Theory and Practice

Chombart de Lauwe, Paul. *Photographies Aériennes*. Paris: 1951. Some excellent examples of the way in which low-level photography can throw light on a whole range of historical problems.

Chombart de Lauwe, Paul, ed. *La Découverte Aérienne du Monde*. Paris: 1948. A collaborative work in which the pioneers of aerial photography in France set out their visions of how the field should develop; splendid and unusual photographs.

Crawford, O. G. S. *Wessex from the Air*. Oxford: 1928. One of the classics; Crawford had first appreciated the potential of aerial photography while serving in the British Army as an aerial observer, and went on to apply the new technique with great success to many problems in British archeology.

Deuel, Leo. *Flights into Yesterday*. New York: 1969. A good illustrated summary of the early years of aerial photography; particularly interesting for its New World material.

Goddard, George W. *Overview: a Life-Long Adventure in Aerial Photography*. New York: 1969. A personal memoir on the development of low-level aerial photography in the United States.

Lee, Willis T. *The Face of the Earth as Seen from the Air*. New York: 1922. An early and now naive-seeming account of the view from the air.

Newhall, Beaumont. *Airborne Camera: The World from the Air and Outer Space*. New York: 1969. An attempt to give a general history of aerial images to 1969; some very striking photographs.

Poidebard, A. "Méthode Aérienne de Recherches en Géographie Historique." *T.A.M., La Géographie* LVII (1932), pp. 1–16. Reflections by one of the early French masters of the field.

Some Recent Examples from Britain and France

Bazzana, A., and A. Humbert. *Prospections Aériennes: Les Paysages et Leur Histoire*. Paris: 1983. This series of articles gives examples of the way low-level oblique photographs have been used in Spain, mostly by French researchers. The photographs are exceptionally clear and the accompanying maps very instructive.

Beresford, M. W., and J. K. S. Saint-Joseph. *Medieval England, an Aerial Survey*. Cambridge: 1958. This book uses the work of the aerial survey unit established at Cambridge University to illuminate many monastic and other sites; Beresford often used aerial photographs in an exemplary way in his many books.

Bradford, John. *Ancient Landscapes: Studies in Field Archeology*. London: 1957. Excellent analysis of the uses of aerial photography for the archeologist, with some reflections as well for the historian of early modern Europe.

Chevallier, Raymond. *La Photographie Aérienne*. Paris: 1971. A rather dull but full summary of the field.

————, ed. *Archéologie Aérienne et Techniques Complémentaires*. Paris: 1963. An exhibition catalog showing the wide range of application of low-level photographs to historical problems.

————, ed. *Mélanges d'Archéologie et d'Histoire Offerts à André Piganiol*. 3 vols. Paris: 1966. Contributions from many leading European practitioners such as Roger Agache and Giulio Schmiedt; good plates and maps.

Gascar, Pierre. *France from the Air*. New York: 1972. An album of beautifully reproduced low-level oblique photographs, without very much supporting text.

Muir, Richard. *History from the Air*. London: 1983. A selection from the collection of the unit at the University of Cambridge; strong both on medieval and early modern times.

Norman, Edward R., and J. K. S. Saint-Joseph. *The Early Development of Irish Society: The Evidence of Aerial Photography*. London: 1969. In this book, Saint-Joseph teamed up with an Irish historian to provide striking images of the early period of Irish history.

Platt, Colin. *Medieval Britain from the Air*. London: 1984.

Saint-Joseph, J. K. S. *The Uses of Air Photography*. New York: 1966.

Stanley, Christopher. *The History of Britain: An Aerial View*. New York/London: 1984. The author was an archeological field officer who used a light aircraft to take his own photographs. The book is exceptionally strong on early earthworks, but has good examples from the whole historical range.

Wilson, D. R. *Air Photo Interpretation for Archaeologists*. New York: 1982.

Some Recent Examples from the United States

In this country, there is nothing to compare with the work that has been done in Europe. The only journal devoted to the field, *Aerial Archeology,* used to have summaries of work done in each region, and the summaries for the United States were invariably scanty. This is a great pity, for the opportunities to do fruitful and revealing work are at least as great here as in Europe. This lack of interest is reflected in the small number of authors to whom we can refer.

Cameron, Robert. This author has compiled *Above London* (San Francisco: 1988); *Above San Francisco* (San Francisco: 1976); *Above Washington* (San Francisco: 1980); and *Above Paris* (San Francisco: 1984). The photographs are often very good, but the text is in general weak and the problems of history are dealt with only peripherally.

Garnett, William. *The Extraordinary Landscape: Aerial Photographs of America*. Boston: 1982. This is a book of extraordinary beauty, but the views have naturally been chosen for their aesthetic rather than for their historical interest.

Gerster, Georg. *Flights of Discovery; the Earth from Above*. New York: 1978. Fine, low-level oblique and vertical photographs from all over the world.

Gleason, David. This author has so far produced *Over New Orleans* (Baton Rouge: 1983) and *Over Boston* (Baton Rouge: 1985). The latter has many interesting views, and the former is a very fine book, no doubt because the author is intimately acquainted with the material, and so knew not only what sites to include, but also what questions to ask of them.

Plate 12.2. Detail of Chicago from the *Satellite Image Map of Northeastern Illinois,* by Richard Dahlberg and Donald Luman, 1985. Illinois State Geological Survey, and Northern Illinois University Laboratory for Cartography and Spatial Analysis.

David Buisseret and Christopher Baruth

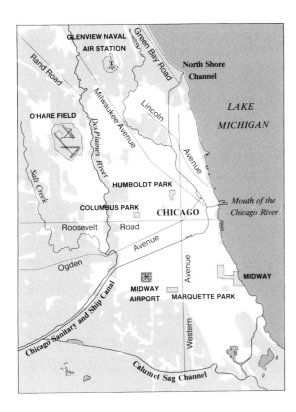

Plate 12.2a. Diagram
of Plate 12.2.

Chicago: The Analysis of an Urban Conglomeration

This plate is a multispectral view of Chicago, extending about twenty miles out from the city center. Here we see a black-and-white image, but the original, in the color section (Color Plate 7), is in false color, with cold areas such as the lake in blackish blue, and the living vegetation giving a red response. Perhaps the most salient features are the three large airfields (Midway, O'Hare, and Glenview), with their conspicuous runways, and the general lines of the hydrography. Here the dark straight lines of the canals are conspicuous in the south, as is the Des Plaines River in the north. Once we have identified the signature of the Des Plaines River, it becomes easier to pick out the line of the North Branch of the Chicago River, and of the North Shore Channel.

Having thus identified the main water features, which are indeed responsible for Chicago's prominence, we can pass to other elements. The checkerboard pattern of the streets is evident, if rather faint. The red response of the parks and cemeteries, whose massed trees and grass give very distinctive signatures, stands out clearly. Note in particular the shapes of the midway (by the University of Chicago) and of Humboldt Park.

Passing on to the communications, the railroad tracks are too narrow to show at this scale, but some roads are very conspicuous. Around O'Hare, in particular, the winding white bands of the recent expressways are very prominent. If we look closely we can also detect the radial roads that have been in position almost from Chicago's birth. In the north, Rand Road, Milwaukee Avenue, Lincoln Avenue, and Green Bay Road were all either plank roads or Indian trails or both; to the south, Ogden Avenue and Western Avenue were the early approach roads. This cursory analysis far from exhausts the interest of this image of Chicago, which we include as an example of the way in which remote sensing allows us to dissect an urban agglomeration.

Source: Satellite Image Map of Northeastern Illinois, *ed. Richard E. Dahlberg and Donald E. Luman (Champaign: 1985).*

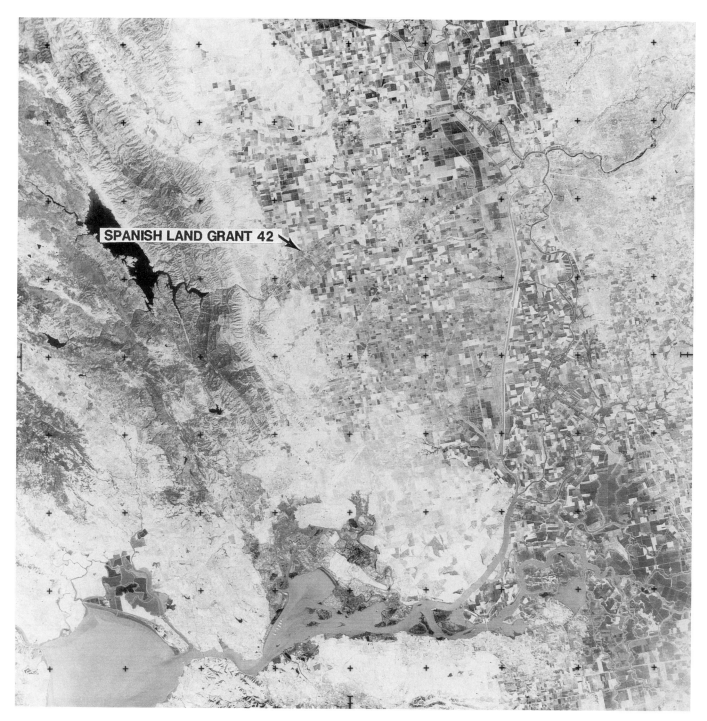

Plate 12.3. LANDSAT image of part of California, June 1978. The American Geographical Society Collection.

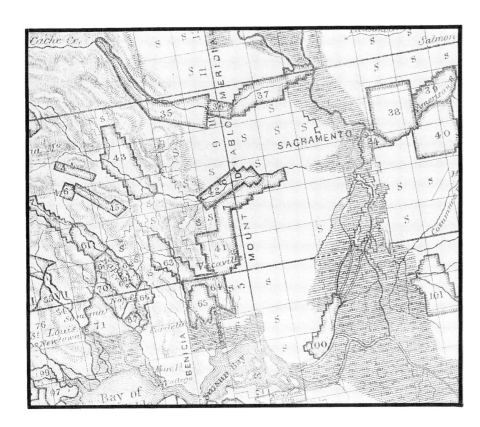

Plate 12.3a. Detail from the *Map of Public Surveys in California* to accompany the report of the Surveyor General (Washington, 1861).

Spanish Land Grants in California's Central Valley

This is an image from LANDSAT III, available only in monochrome but with excellent spatial resolution, or high detail. It has been chosen to show the patchwork of cadastral systems found in California's Central Valley. This valley is a flat, low-lying plain drained in the north by the Sacramento River and in the south by the San Joaquin. As the detail from the 1861 map shows, the lands adjacent to the rivers were swampy, and not surveyed. The map also shows the rectangular survey of the General Land Office in juxtaposition to the old Spanish land grants, of which no fewer than 411 are listed. The Spanish land grants tended to be very large, since the country was often devoted to ranching, and to have roughly rectangular shapes, unlike the Anglo metes and bounds of the east (see for instance Plates 3.4 and 3.5).

The LANDSAT image shows the effect on the landscape of both these, but also that the swamps have been drained and the former swampland divided into fields. This subdivision was not in accord with the rectangular survey, but in places rather reminiscent of the French long lots seen on Plate 4.6. This whole area between Sacramento and San Francisco has of course undergone intense development over the past century, so that what was virtually empty countryside is now a patchwork of fields, cities, roads, railroads, and canals. Even so, it is still hemmed in to the west by mountains, upon which the human impact is not so evident.

Plate 12.4. LANDSAT image of Green Bay, Wisconsin, September 1982. The American Geographical Society Collection.

David Buisseret and Christopher Baruth

Landholding at
Green Bay, Wisconsin

Here is a detail from an image captured by the type of multispectral scanner aboard LAND-SATS IV and V. This machine could produce very versatile false-color presentations, and in black and white permitted an attention to detail impossible before the late 1970s. In this enlargement, for instance, it is possible to see all of the streets in Green Bay, Wisconsin.

This city traces its origin to the French fur trappers of the seventeenth century, and, as in other sites shown elsewhere in this book (Plates 3.6 and 4.6), has their distinctive long-lot system of land tenure. From Fort Howard, at the mouth of the Fox River, south to the rapids, these long lots lined the river, as the 1835 map shows. By then, though, the rectangular divisions of the General Land Office had reached Wisconsin, and the new squares were carefully fitted in with the old strips, something that was essential if preexisting tenurial rights were to be respected in the new Republic.

This compromise between two systems is very obvious on the image taken in 1982. Here we are struck not only by the way in which the long-lot orientation has determined the main lines of the town, but also by the uniformity with which roads have sprung up along the division lines of the General Land Office system. These roads, outlined here in white, also allow us to see with what precision the surveyors of the 1830s worked, little expecting that one day their work would be subject to aerial inspection!

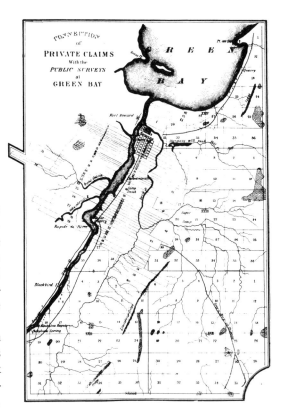

Plate 12.4a. *Connection of Private Claims with the Public Surveys at Green Bay,* compiled from Public Surveys as Returned to the Surveyor General's Office . . . (Washington, 1935).

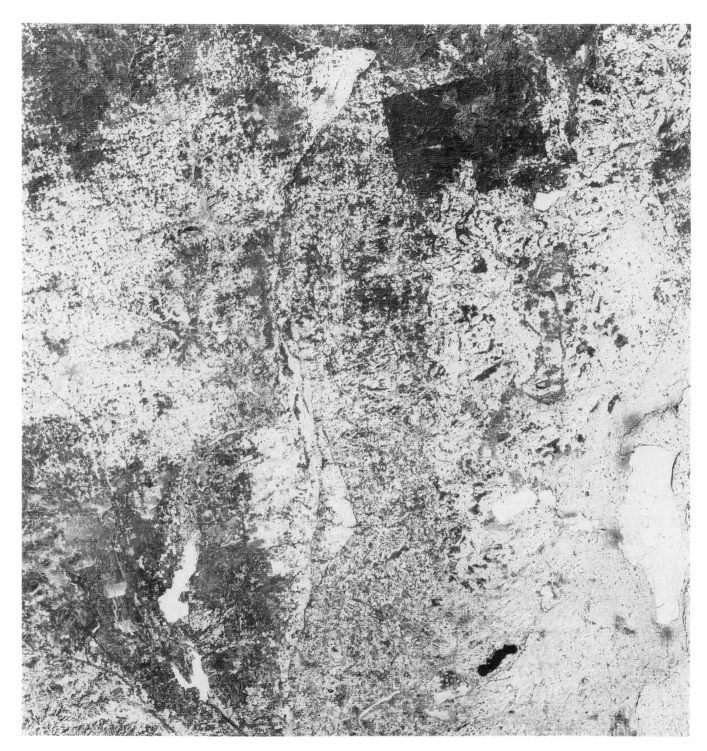

Plate 12.5. LANDSAT image of east-central Wisconsin, January 1976. The American Geographical Society Collection.

David Buisseret and Christopher Baruth

Differential Forest Cover on the Menominee Indian Reservation, Wisconsin

This is an image of snow-covered central Wisconsin taken January 4th, 1976. Near the upper right-hand corner is a dark rectangular area that is notable for the linearity of its bounds, and for the distinctive shape of its southern edge. Consulting Plate 12.5a, we see that this dark area corresponds to Menominee County, home of the Indians of that tribe, and to the adjacent Stockbridge Indian Reservation.

The reason for the contrast between the appearance of the area settled by Indians and that of the neighboring countryside lies in comparative land use. While most surrounding lands have been cleared for agriculture, the Indian lands have been retained as cultivated forest. In this bleak winter scene, central Wisconsin is snow-covered and ice-bound. LANDSAT readily images the snow where it covers the fields, but not where hidden beneath dense woodland. The result is that the county and reservation stand in sharp distinction to the rest. This contrast is maintained throughout the year, though intensified at some seasons. For instance, during the growing season the purplish hue of the coniferous reservation forest stands out sharply from the red of the surrounding cultivated fields.

The reservation's forest contrasts not only with the cultivated fields, but also with other forests. Partially abutting the northern boundary of the reservation and extending along most of the top edge of the scene is the southern limit of Wisconsin's northern forest lands. This mixed forest, occupying lands not suitable for agriculture, appears decidedly lighter in tone than the reservation.

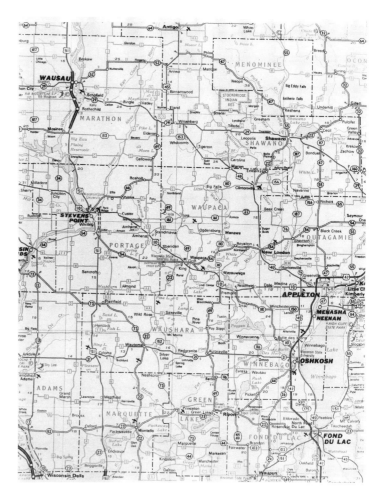

Plate 12.5a. Detail of east-central Wisconsin, from *Wisconsin* 1969. Road map issued by the Shell Oil Company.

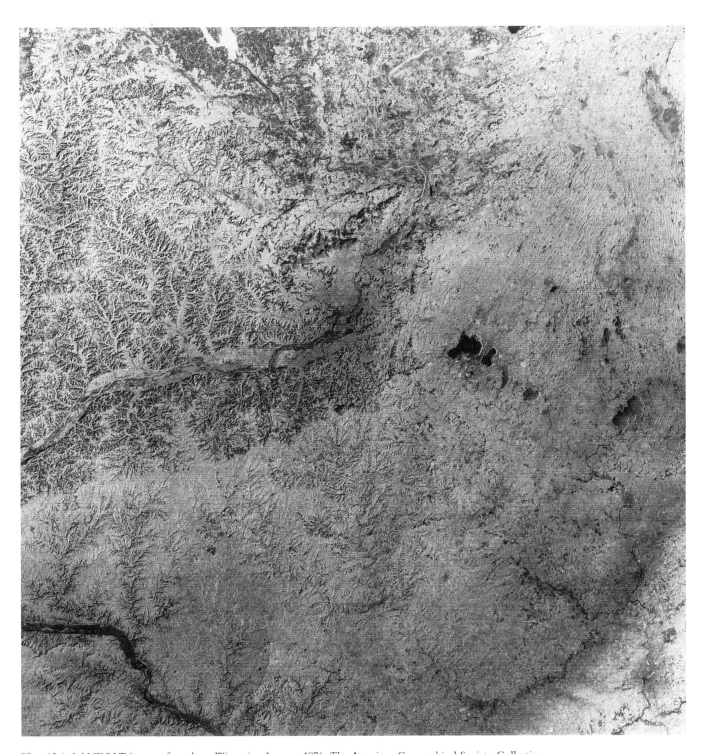

Plate 12.6. LANDSAT image of southern Wisconisn, January 1976. The American Geographical Society Collection.

Geological Features Relating to Patterns of Settlement

This scene covers much of southwestern Wisconsin, and though it is adjacent to the previous site, reveals a distinctly different landscape. This one has largely resisted the settlers' geometric designs, and has indeed dictated the routes followed by the region's transportation network; the inevitable town line road has given way to random thoroughfares that conform to valley or crest.

One such crest is "Military Ridge," extending from just west of Madison (located between the two open lakes just right of center) to the Mississippi River. Military Ridge, as is apparent from our detail of the Raisz landform map, is a drainage divide of striking linearity that separates the streams flowing north into the Wisconsin River from those flowing southward into the Mississippi. The well-integrated dendritic or tree-like pattern of the drainage system is the consequence of the absence of continental glaciation in the region.

Military Ridge received its name when the Army built a supply road there in 1832 during the Black Hawk War. It was the perfect place for a road, for it was reasonably level, and as it is a drainage divide, no rivers needed to be bridged. It also is high and dry with no danger of flooding. Because of these natural advantages, the ridge was later suggested as a railroad route, but the railroad was ultimately built on the floodplain of the nearby Wisconsin River, whose flatness was too great an asset to ignore.

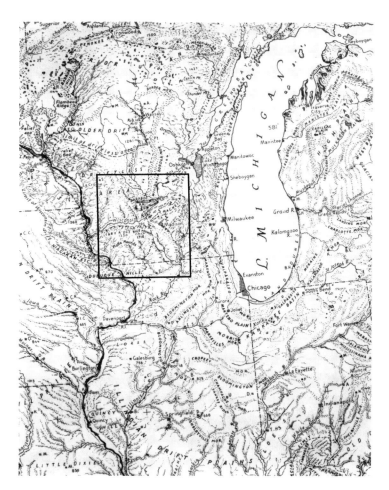

Plate 12.6a. The area of Plate 12.6, in *Landforms of the United States* by Erwin Raisz (6th edition, 1957).

Plate 12.7. Aerial view of Indian earthwork, 1934. The Smithsonian Institution, Dache M. Reeves Collection.

David Buisseret and Christopher Baruth

An Example of a Geometrical Indian Earthwork

This medium-level aerial photograph, taken in 1934, shows the extraordinary earthwork still preserved just to the south of Raccoon Creek, about a mile to the west of Newark, Ohio. This earthwork, which stands out with remarkable clarity on the photograph, is marked on the USGS map as "Octagon State Memorial." It testifies not only to the labor available to the Indians of this region, but also, in the precision of its geometrical forms, to the advanced level of their surveying skills. Earthworks of this kind can, of course, be surveyed and plotted on the ground. But the aerial view instantly gives a wider and more accurate image than would be possible except after months of labor on the ground.

As might be expected, the area around the monument has been developed to a considerable extent since 1934, when the photograph was taken, and 1952, when the map was made. The outline of the roads remains the same, but there are now many houses where there once were open fields. In looking at a photograph of this kind, one cannot help but reflect that many less spectacular sites must have perished, and must still be perishing, under the tractors of the developers. It would seem that future damage might yet be minimized by the use of aerial photographs to record such sites.

Source: William N. Morgan, Prehistoric Architecture in the Eastern United States *(Cambridge/London: 1980).*

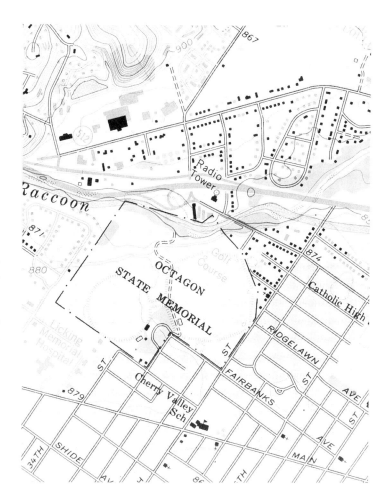

Plate 12.7a. Detail from USGS 7.5-minute quad for Newark, Ohio (1952 edition).

Plate 12.8. Aerial view of
Destrehan, on the Mississippi
River. Gleason Photography,
Baton Rouge.

David Buisseret and Christopher Baruth

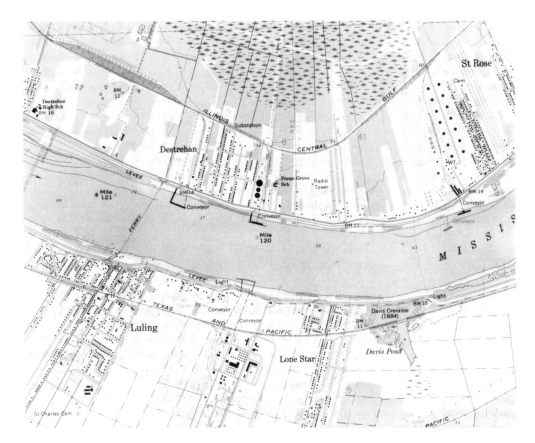

Plate 12.8a. Detail from USGS 7.5-minute quads for Luling and Hahnville, Louisiana (1953 and 1958 editions).

Studying the Location of a Louisiana Plantation House

This photograph was taken looking southeast, from a point roughly above "Destrehan High School" on the USGS map. In the foreground is the plantation house of Destrehan, a fine example of its type, and in the background are the industrial installations with their two wharves, each marked "Conveyor" on the map. Along the middle of the photograph runs the River Road, which in this part of the world almost always separates the plantations and their houses from the levee and the Mississippi River. The USGS map (Plate 12.9a) offers a good example of this general layout, where most of the plantations have their buildings near the river, the fields stretching behind them back into the bush.

On the ground, it is impossible to get any idea of the surroundings of Destrehan, because the levee completely cuts off the view to the river side, and the vegetation obscures the view inland. With the aerial photograph, we can at once seize the relationship between the house, the plantation, and the river. We also have a most graphic image of the way in which this part of Louisiana has been transformed, from a primarily agricultural sugar-producing economy into one dominated by huge industrial enterprises, based on the great river.

Source: David King Gleason, Over New Orleans *(Baton Rouge/London: 1983).*

Plate 12.9. Aerial view of
Laurel Valley Plantation, near
Thibodaux, Louisiana.
Gleason Photography, Baton
Rouge.

David Buisseret and Christopher Baruth

Plate 12.9a. Detail from USGS 7.5-minute quad for Thibodaux (1961 edition).

The Disposition of Slave Quarters on a Plantation

Laurel Valley was a very large sugar estate, founded in 1840, and its remaining buildings form the largest such surviving complex. As the USGS map (Plate 12.9a) shows, it lies some way outside the town of Thibodaux, inland on one of the old French long lots. We may wonder in passing if its inland siting is the result of its relatively late foundation, since on the whole such establishments tended in this part of the world to lie alongside the river, as we have seen.

In the photograph, the factory is at the middle right, in ruins after hurricane Betsy of 1965. The structure and disposition of the slave cabins are very striking, seen from the air. The cabins are uniform, squarely built, and arranged in rows. This aerial image shows us a classic "European-type" slave quarters, with living arrangements at the opposite pole from the "African-type" quarters once found in the Caribbean. There, the huts were round, and irregularly placed among the trees, which offered both shade and food. The aerial photograph, on the contrary, conveys a strong feeling for the "industrial," Europeanized type of quarters, with much more feeling for order and logic than for the pleasure or comfort of the inhabitants.

Source: David King Gleason, Over New Orleans *(Baton Rouge/London: 1983).*

Plate 12.10. Aerial view of
Fort Union, New Mexico.
William Garnett.

David Buisseret and Christopher Baruth

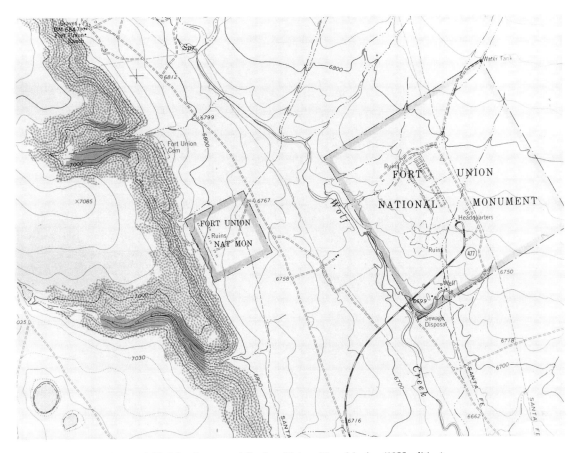

Plate 12.10a. Detail from USGS 7.5-minute quad for Fort Union, New Mexico (1955 edition).

An Image of Europe in the American Desert

This remarkable photograph, which is even more striking in color (Color Plate 8), was taken late one evening by William Garnett, the Western photographer. The star-shaped Fort Union, dating to 1863–1866, is caught in the rays of the sinking sun with a detail that could never be achieved through ground survey. The photograph also brings out very well the stark nature of the country in which the fort was established, not only as a defence against the Confederate forces, but also as part of a campaign to subjugate the local Indian tribes.

In this harsh climate, wagon tracks survive almost indefinitely. The ones leading away to the north, at the top middle of the photograph, arrive eventually at Denver; the ones leading off to the right are the continuation of the Santa Fe Trail, which soon reaches the settlement of that name. This image of Fort Union is a fine example of the way in which an aerial photograph can bring out the very bones of an historic site.

Source: James Arrott, Arrott's Brief History of Fort Union *(Las Vegas: 1962); and William Garnett,* The Extraordinary Landscape: Aerial Photographs of America *(Boston: 1982).*

Plate 12.11. Aerial view of
construction of Burnham
Harbor, Chicago. American
City Views.

One Stage in a Great Building Project: Burnham Harbor, Chicago

This marvelously sharp photograph, taken in 1930, shows what would become Burnham Park Harbor in Chicago, at a time when the land for it was being taken in from the lake. Notice the rounded promontory on which the Adler Planetarium would soon stand, and the curving line of each side of the harbor behind it. Notice, too, the huge structure of the Field Museum of Natural History, with the Soldier Field Stadium taking shape to the south of it. At Soldier Field, the southern semicircle of buildings is as yet only marked out on the ground.

This photograph is a good example of the ability of aerial photographs to give instant, accurate information on transient phenomena of considerable scale. Of course, this information could range from continentwide communications systems to very detailed images of buildings under construction, or of the flow of vehicles along a route. The advantage of the aerial photograph to the historian, and indeed to the military analyst among others, is that these artifacts are frozen at one instant in time, and then available for later leisurely analysis.

Source: George W. Goddard, Overview: a Life-Long Adventure in Aerial Photography *(New York: 1969).*

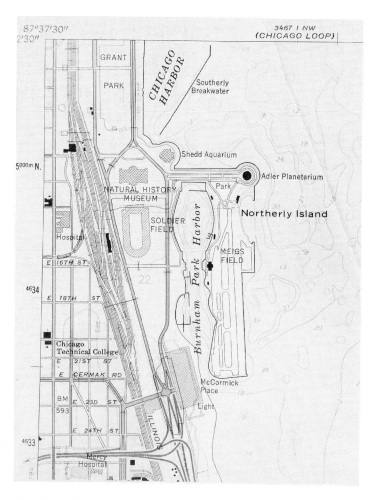

Plate 12.11a. Detail from USGS 7.5-minute quad of Jackson Park, Illinois (1972 edition).

Sources of General Land Office Maps

A. Addresses of Bureau of Land Management (BLM) offices holding collections for more than one state.
  1. BLM–ESO:
     BLM, Eastern States Office
     350 South Pickett Street
     Alexandria, Virginia 22304
     703-235-2861
  2. BLM–Montana:
     BLM, Montana State Office
     222 North 32nd Street
     P.O. Box 30157
     Billings, Montana 59107
     406-657-6090
  3. BLM–Oregon:
     BLM, Oregon State Office
     825 NE Multnomah
     P.O. Box 2965
     Portland, Oregon 97208
     503-231-6273
B. Addresses of National Archives and Records Administration (NARA) offices holding collections for more than one state.
  1. NARA–NNSC:
     National Archives, Cartographic and
        Architectural Branch
     841 South Pickett Street
     Alexandria, Virginia 22304
     703-756-6700
  2. NARA–Denver:
     National Archives, Denver Branch
     Building 48, Denver Federal Center
     Denver, Colorado 80225
     303-236-0817

  3. NARA–Denver (FRC):
     Denver Federal Records Center
     Building 48, Denver Federal Center
     Denver, Colorado 80225
     303-236-0804
  4. NARA–San Francisco:
     National Archives, San Francisco Branch
     1000 Commodore Drive
     San Bruno, California 94066
     415-876-9009
  5. NARA–Seattle:
     National Archives, Seattle Branch
     6125 Sand Point Way NE
     Seattle, Washington 98115
     206-526-6507
  6. NARA–Seattle (FRC):
     Seattle Federal Records Center
     6125 Sand Point Way NE
     Seattle, Washington 98115
     206-526-6501
C. Disposition of plats and field notes by state.
  1. Alabama
     a. Original paper records
        (1) Original paper records
            Alabama Department of Archives and
               History
            624 Washington Avenue
            Montgomery, Alabama 36130
            205-261-4361
        (2) Microfilm copy
            Secretary of State
            Lands and Trademarks Division
            528 State Office Building

Montgomery, Alabama 36104
205-261-5325
   b. Duplicate plat and field notes
      BLM–ESO
   c. Triplicate plat
      NARA–NNSC
2. Alaska
   a. Original plat and field notes
      BLM, Alaska State Office
      701 "C" Street, Box 13
      Anchorage, Alaska 99513
      907-271-5555
   b. Duplicate plat and field notes
      BLM–ESO
   c. Triplicate plat
      NARA–Seattle (FRC)
3. Arizona
   a. Original plats and field notes
      BLM, Arizona State Office
      2400 Valley Bank Center
      Phoenix, Arizona 85073
      602-241-5544
   b. Duplicate plats and field notes
      BLM–ESO
   c. Triplicate plats
      NARA–Denver (FRC)
4. Arkansas
   a. Original plats and field notes, triplicate plats
      (incomplete)
      Arkansas State Land Office
      State Capitol, Room 109
      Little Rock, Arkansas 72201
      501-371-1222
   b. Duplicate plats and field notes
      BLM–ESO
5. California
   a. Original plats and field notes
      BLM, California State Office
      Federal Office Building
      2800 Cottage Way
      Sacramento, California 95825
      916-484-4014
      (Original records prior to 1906 were
      destroyed in San Francisco fire and
      earthquake. This office has original records
      created after 1906, as well as lithographic
      copy of the duplicate plats maintained by
      BLM–ESO.)
   b. Duplicate plats and field notes
      BLM–ESO
   c. Triplicate plats
      NARA–San Francisco
   d. A fourth copy is maintained by

California State Lands Commission
1807 13th Street
Sacramento, California 95814
916-322-3589
6. Colorado
   a. Original and triplicate plats
      NARA–Denver
   b. Original field notes and microfilm copies of
      plats
      BLM, Colorado State Office
      1037 20th Street
      Denver, Colorado 80202
      303-294-7627
   c. Duplicate plats and field notes
      BLM–ESO
7. Florida
   a. Original plats and field notes
      Florida Department of Natural
         Resources
      Division of State Lands, Bureau of State
         Lands Management, Title Section
      3900 Commonwealth Boulevard
      Tallahassee, Forida 32303
      904-488-8123
   b. Duplicate plats and field notes
      BLM–ESO
   c. Triplicate plats
      Location unknown
8. Idaho
   a. Original plats and field notes, triplicate plats
      BLM, Idaho State Office
      Federal Building
      550 West Fort Street
      P.O. Box 042
      Boise, Idaho 83724
      208-334-1438
   b. Duplicate plats and field notes
      BLM–ESO
9. Illinois
   a. Original plats and field notes
      Illinois State Archives
      Springfield, Illinois 62756
      217-782-4682
   b. Duplicate plats and field notes, triplicate plats
      NARA–NNSC
10. Indiana
   a. Original plats and field notes
      Indiana Commission on Public
         Records, Archives Division
      State Library Building, Room 117
      140 North Senate Avenue
      Indianapolis, Indiana 46204
      317-232-3660

b. Duplicate plats and field notes, triplicate plats
NARA–NNSC

11. Iowa
    a. Original plats and field notes
       Iowa State Archives
       Records and Property Center,
          4th Floor
       East 7th and Court Avenue
       Des Moines, Iowa 50319
       515-281-3007
    b. Duplicate plats and field notes, triplicate plats
       NARA–NNSC

12. Kansas
    a. Original plats and field notes
       (1) Original plats
           Kansas Secretary of State
           State House, 2nd Floor
           Topeka, Kansas 66612
           913-296-2236
       (2) Original field notes and microfilm copy of
           plats
           Kansas State Historical Society
           120 West Tenth Street
           Topeka, Kansas 66612
           913-296-3251
    b. Duplicate plate and field notes, triplicate plats
       NARA–NNSC

13. Louisiana
    a. Original plats and field notes, triplicate plats
       Louisiana Department of Natural
          Resources
       Division of State Lands
       P.O. Box 44124
       Baton Rouge, Louisiana 70804
       504-342-4577
    b. Duplicate plats and field notes
       BLM–ESO

14. Michigan
    a. Original plats and field notes
       Michigan Department of Natural
          Resources
       Lands Division, Land Records
       Box 30028
       Lansing, Michigan 48926
       517-373-1250
    b. Duplicate plats and field notes
       BLM–ESO
    c. Triplicate plats (Complete set not available:
       Some triplicates as well as partial set of 1894
       lithographic copies are available at the follow-
       ing two repositories)
       (1) State Archives of Michigan
           3405 North Logan

Lansing, Michigan 48918
517-373-0512
(2) Bentley Historical Library
    University of Michigan
    1150 Beal Avenue
    Ann Arbor, Michigan 48109
    313-764-3482

15. Minnesota
    a. Original plats and field notes
       Secretary of State
       Foreign Corporations Section
       180 State Office Building
       St. Paul, Minnesota 55155
       612-296-9215
    b. Duplicate plats and field notes
       BLM–ESO
    c. Triplicate plats
       Department of Natural Resources
       Bureau of Land
       500 Lafayette Road
       St. Paul, Minnesota 55146
       612-296-0659

16. Mississippi
    a. Original plats and field notes
       Secretary of State
       Public Lands Division
       P.O. Box 136
       Jackson, Mississippi 39205
       601-359-1383
    b. Duplicate plats and field notes
       BLM–ESO
    c. Triplicate plats
       NARA–NNSC

17. Missouri
    a. Original plats and field notes
       Department of Natural Resources
       Division of Geology and Land Survey
       Land Survey Program
       P.O. Box 250
       Rolla, Missouri 65401
       314-364-1752
    b. Duplicate plats and field notes, triplicate plats
       NARA–NNSC

18. Montana
    a. Original plats
       NARA–Denver
    b. Duplicate plats and field notes
       BLM–ESO
    c. Triplicate plats, microfiche copy of field notes
       BLM–Montana

19. Nebraska
    a. Original plats and field notes
       Nebraska State Surveyor

301 Centennial Mall, South
P.O. Box 94663, State Office Building
Lincoln, Nebraska 68509
402-471-2566

b. Duplicate plats and field notes
BLM–ESO

c. Triplicate plats
Nebraska State Historical Society
1500 R Street
Lincoln, Nebraska 68508
402-432-2793

20. Nevada
a. Original plats and field notes
(1) Original plats
NARA–San Francisco
(2) Recent plats, original field notes, microfilm
copy of plats
BLM, Nevada State Office
Federal Building, Room 3008
300 Booth Street
Reno, Nevada 89509
702-784-5787

b. Duplicate plats and field notes
BLM–ESO

c. Triplicate plats
NARA–San Francisco

21. New Mexico
a. Original plats and field notes, triplicate plats
BLM, New Mexico State Office
U.S. Post Office and Federal Building
South Federal Place
P.O. Box 1449
Santa Fe, New Mexico 87501
505-988-6316

b. Duplicate plats and field notes
·BLM–ESO

22. North Dakota
a. Original plats and field notes
State Water Commission
State Office Building
900 East Boulevard
Bismarck, North Dakota 58505
701-224-2750

b. Duplicate plats and field notes
BLM–ESO

c. Triplicate plats
BLM-Montana

23. Ohio
a. Original plats and field notes
Ohio State Auditor's Office
Land Office
88 East Broad Street
Columbus, Ohio 43215
614-466-6206

b. Duplicate plats and field notes, triplicate plats
NARA–NNSC

24. Oklahoma
a. Original plats and field notes
BLM–ESO

b. Triplicate plats (partial duplicate and triplicate
sets for Oklahoma Territory only)
NARA–NNSC

25. Oregon
a. Original plats and field notes
BLM–Oregon

b. Duplicate plats and field notes
BLM–ESO

c. Triplicate plats
NARA–Seattle

26. South Dakota
a. Original plats and field notes
(1) Original paper records
State Library and Archives
800 North Illinois Street
Pierre, South Dakota 57501
605-773-3173
(2) Microfilm copies of plats and field notes
School and Public Land Commission
State Capitol
Pierre, South Dakota 57501
605-773-3303

b. Duplicate plats and field notes
BLM–ESO

c. Triplicate plats
BLM–Montana

27. Utah
a. Original plats and field notes
BLM, Utah State Office
University Club Building
136 East South Temple
Salt Lake City, Utah 84111
801-524-3038

b. Duplicate plats and field notes
BLM–ESO

c. Triplicate plats
NARA–Denver

28. Washington
a. Original plats and field notes
BLM–Oregon

b. Duplicate plats and field notes
BLM–ESO

c. Triplicate plats
NARA–Seattle

29. Wisconsin
a. Original plats and field notes
(1) Original records
Board of Commissioners of Public Lands
505 Segoa Road

Madison, Wisconsin 53705
608-266-1370
(2) Manuscript, photostatic, and microfilm
copies
State Historical Society of Wisconsin
Archives Division
816 State Street
Madison, Wisconsin 53706
608-262-3338
b. Duplicate plats and field notes
BLM–ESO
c. Triplicate plats
NARA–NNSC

30. Wyoming
a. Original plats
NARA–Denver
b. Duplicate plats and field notes
BLM–ESO
c. Triplicate plats, original field notes
BLM, Wyoming State Office
2515 Warren Avenue
P.O. Box 1828
Cheyenne, Wyoming 82001
307-772-2334

Glossary

Atlas a collection of maps in a volume. This usage derives from the god Atlas, who was supposed to hold up the pillars of the universe; his name was used for Mercator's collection of maps.

Base line one of the east-west lines off which the township and range system is calculated.

Base map a map showing only the basic geographical features of an area, often used for plotting the type of material that will eventually form a thematic map.

Benchmark relatively permanent material object, natural or artificial, whose location and elevation are accurately known; from this point further measurements can be made with confidence.

Bird's-eye view image of a geographical area, often a town, taken from a high oblique angle.

Cadastral map map relating to land boundaries, usually drawn in order to define the limitations of title.

Cardinal points north, south, east, and west.

Cartography the making of maps, plans, and charts.

Chart a map normally designed for navigation, whether by sea or air or in the heavens.

Chorography the mapping of an area of regional size, as opposed to geography and topography.

Compass can confusingly mean either an instrument for describing circles or a device for indicating magnetic north.

Contour a line on a map linking points that are of equal altitude on the ground.

Coordinates sets of figures designating a certain point in relation to some given frame of reference, for example, a grid.

Elevation vertical distance of a point above or below some reference surface, such as the level of the sea.

Engraving way of making a map by printing from an inscribed metal plate; maps can also be made, for instance, by hand (manuscript) or by lithography.

Geography the science of describing the earth's surface, formerly used in contrast to chorography, which applied to some limited area of the same.

Grid network of uniformly spaced parallel lines intersecting at right angles; used on maps for establishing coordinates.

Hachures lines used on a map to indicate the direction and steepness of slopes.

Hydrography the science of describing the physical features of watercourses and their adjacent land.

LANDSAT abbreviation for a series of unmanned earth-orbiting NASA satellites that acquire multispectral images.

Latitude angular distance in degrees, minutes, and seconds of a point north or south of the equator (from 0 to 90 degrees).

Lithograph an image made by using inks and waxes on a special stone.

Longitude angular distance in degrees, minutes, and seconds of a point east or west of the Greenwich meridian (from 0 to 360 degrees; formerly a variety of other meridians were used, for example, that of the Azores).

Map a graphic representation of all or part of the earth's surface; sometimes this definition may be much more restrictive, taking in such elements as scale or orientation.

Meridian one of an infinite series of imaginary north-south lines passing through both poles. All points on a given meridian have the same longitude.

Multispectral image image (of the earth's surface) derived from scanners sensing several channels of the electromagnetic spectrum (visible light, infrared waves, etc.).

Ordnance survey (of Great Britain) survey operation, begun in the eighteenth century, by which Great Britain was covered with topographical maps of considerable accuracy and some originality.

Plan graphic description of some geographical area, usually quite limited in extent.

Planimetric map map presenting precise horizontal positions for features represented, as opposed for instance to a bird's-eye view.

Plat a relatively simple map of some restricted area, often drawn by a surveyor in order to define title to land.

Portolan chart map deriving from the medieval portolans, or navigation instructions; these charts are distinctive in style and at first covered only the Mediterranean Sea.

Prime meridian section of a meridian chosen for establishing the east-west measurements of the township and range system.

Profile a view (almost always of a city) taken from ground level, usually with the object on the horizon; to be contrasted with both bird's-eye views and planimetric maps.

Projection the mathematical concept following which the cartographer solves the problem of representing the round earth on a flat surface.

Quads short for quadrangles, the subdivisions into which USGS topographical maps are divided.

Range the east-west component of the township and range system; ranges are measured east or west from a prime meridian.

Relief the elevations or depressions of the land or sea bottom.

Remote sensing process of detecting and describing the properties of an area by measuring its reflected and emitted radiation.

Scale the relationship existing between the distance on a map and the corresponding distance on the earth. Confusingly enough, a small-scale map probably shows a large area, and vice versa.

Strip map map on which information is presented as part of a narrow scroll; such maps are often used to show routes.

Thematic map map designed to provide information on some specialized subject such as rainfall or population distribution; to be distinguished from base maps, which show general geographical features.

Topographical map map showing relief and sometimes vegetation in detail; formerly a map showing a restricted area such as a village or estate, in contrast to chorographical or geographical maps.

Toponym a place-name.

Township the basic element in the township and range system, which groups these six-mile-square units according to their relationship to certain base lines and prime meridians.

USGS maps maps produced by the U.S. Geological Survey; they provide the best topographical coverage for the United States, and form the indispensable base maps for many historical studies.

Wind rose a circle with the directions of the winds inscribed at its margins; these circles are often very decorative, and found on many portolan charts.

Woodcut a print derived from an impression cut into a block of wood, as opposed to an engraving.

Index

68968

DATE DUE

MAR 2 2 2001			
MAR 1 2 2001			

AUGUSTANA UNIVERSITY COLLEGE
LIBRARY